DR
SCHOOL B
QUEEN MARY COLLEGE,
MILE END ROAD,
LONDON E1 4NS
01-980-4811 x 4160

The Phototrophic Bacteria:
Anaerobic Life in the Light

Studies in Microbiology

EDITORS

N. G. CARR
Department of Biochemistry
University of Liverpool

J. L. INGRAHAM
Department of Bacteriology
University of California
at Davis

S. C. RITTENBERG
Department of Bacteriology
University of California
at Los Angeles

Studies in Microbiology Volume 4

The Phototrophic Bacteria:
Anaerobic Life in the Light

EDITED BY

J. G. ORMEROD
M.Sc, PhD
Associate Professor
University of Oslo

BLACKWELL SCIENTIFIC PUBLICATIONS
OXFORD LONDON EDINBURGH
BOSTON MELBOURNE

© 1983 by
Blackwell Scientific Publications
Editorial offices:
Osney Mead, Oxford, OX2 OEL
8 John Street, London, WC1N 2ES
9 Forrest Road, Edinburgh, EH1 2QH
52 Beacon Street, Boston,
 Massachusetts 02108, USA
99 Barry Street, Carlton,
 Victoria 3053, Australia

All rights reserved. No part of this publication
may be reproduced, stored in a retrieval system,
or transmitted, in any form or by any means,
electronic, mechanical, photocopying, recording
or otherwise without the prior permission
of the copyright owner

First published 1983

British Library
Cataloguing in Publication Data

The Phototrophic bacteria.—(Studies in
 microbiology; v. 4)
 1. Bacteria   2. Light—Physiological effect
 I. Ormerod, J. G.   II. Series
 576'.6482   QH545.L/

ISBN 0-632-00783-4

Distributed in the United States and Canada by
University of California Press
Berkeley, California

Set by Macmillan India Ltd., Bangalore
Printed by The Alden Press, Oxford and
bound by Butler and Tanner Ltd., Frome and London

Contents

	Contributing authors	ix
	Preface	x
1	Introduction: J. G. ORMEROD	1
2	Structure of phototrophic bacteria; development of the photosynthetic apparatus: J. OELZE	
	2.1 Introduction	8
	2.2 Fine structure of cells and membranes	9
	2.3 Composition and structure of chromatophore vesicles	11
	2.4 Development of the photosynthetic apparatus	17
	2.4.1 Regulation of bacteriochlorophyll synthesis	18
	2.4.2 Formation of reaction centre and light-harvesting bacteriochlorophyll units	20
	2.4.3 Formation of photosynthetically active membranes	23
	2.4.4 Development of the photosynthetic apparatus in synchronously dividing populations of *Rhodopseudomonas sphaeroides*	25
	2.4.5 Development of light-dependent activities	26
	2.5 Summary	29
3	The early photochemical events in bacterial photosynthesis: M. C. W. EVANS AND P. HEATHCOTE	
	3.1 Introduction	35
	3.2 The purple bacteria	35
	3.2.1 What are the photosynthetic pigments in phototrophic bacteria?	35
	3.2.2 How is the light energy trapped?	38
	3.2.3 The photosynthetic unit	39
	3.2.4 Isolation and characterization of the reaction centre	41
	3.2.5 The mechanism of energy conservation in the reaction centre	41
	3.2.6 Identification and properties of the reaction centre components	43
	3.2.7 Structure and function of the reaction centre	50
	3.3 The green phototrophic bacteria	52
	3.3.1 The light-harvesting apparatus	52
	3.3.2 Photosynthetic reaction centres in green bacteria	55
	3.4 Summary	57
4	Production and dissipation of membrane potential; formation of ATP and reducing equivalents: M. C. W. EVANS	
	4.1 Introduction	61
	4.2 The electron transport chain	62

	4.2.1	Components of the *bc* complex	62
	4.2.2	The pathway of electron transfer	64
4.3	ATP synthesis		68
	4.3.1	Proton gradient formation	69
	4.3.2	Membrane potential generation	70
	4.3.3	Conversion of $\Delta^\mu H^+$ into ATP	71
4.4	Low potential reductant generation		71
4.5	Organization of the chromatophore membrane		72

5 Electron donor metabolism in phototrophic bacteria:
T. A. HANSEN

5.1	Introduction		76
5.2	Inorganic and C_1-compounds as photosynthetic electron donors		78
	5.2.1	What substrates are utilized as electron donors and how?	78
	5.2.2	Reduced sulphur compounds as electron donors	80
	5.2.3	Hydrogen as electron donor	87
	5.2.4	Methanol and formate as electron donors	89
5.3	Chemolithotrophy		90
	5.3.1	Reduced sulphur compounds as energy source	90
	5.3.2	Hydrogen, formate and methanol as sources of energy	91
5.4	Anoxygenic photosynthesis of cyanobacteria		92
5.5	Concluding remarks		95

6 Essential aspects of carbon metabolism: J. G. ORMEROD
AND R. SIREVÅG

6.1	Introduction		100
6.2	CO_2 assimilation (in the absence of organic substrates)		101
	6.2.1	The Calvin cycle	102
	6.2.2	The reductive tricarboxylic acid cycle	106
	6.2.3	Concluding remarks	108
6.3	How organic compounds are metabolized		109
	6.3.1	Green sulphur bacteria	109
	6.3.2	The green gliding bacteria	111
	6.3.3	Purple bacteria	111
6.4	Carbon reserve materials and endogenous metabolism		114
	6.4.1	Glycogen	114
	6.4.2	Poly-β-hydroxybutyrate	115
	6.4.3	Breakdown of reserve materials; endogenous metabolism	115

7 Nitrogen fixation and ammonia assimilation: B. C. JOHANSSON,
S. NORDLUND AND H. BALTSCHEFFSKY

7.1	Introduction		120
7.2	Nitrogen fixation		122
	7.2.1	Some physiological aspects of nitrogen fixation	122
	7.2.2	Purification and properties of nitrogenase	123
	7.2.3	Activation of the Fe-protein and metabolic regulation of nitrogen fixation	126
7.3	Ammonia assimilation		130
	7.3.1	Pathways of ammonia assimilation	130
	7.3.2	Purification and molecular properties of glutamine synthetase	132

	7.3.3	Regulation of glutamine synthetase		134
	7.3.4	Biosynthesis of amino acids and its regulation		136
7.4	Genetic studies of nitrogen fixation and ammonia assimilation			137
7.5	Concluding remarks			140

8 Ecology of phototrophic bacteria: H. VAN GEMERDEN AND H. H. BEEFTINK

8.1	Introduction		146
8.2	Field studies		148
	8.2.1	Stratified lakes	149
	8.2.2	Other habitats	154
8.3	Experimental ecology of phototrophic bacteria		154
	8.3.1	Population interactions	155
	8.3.2	Coexistence of competing species	159
	8.3.3	Maintenance and survival	165
	8.3.4	Enrichment cultures	168
8.4	Concluding remarks		171
Appendix: Cultivation methods			179

9 Genetics and molecular biology: B. L. MARRS

9.1	Introduction		186
9.2	Protein synthesis		187
	9.2.1	Messenger RNA	188
	9.2.2	Transfer RNA	189
	9.2.3	Ribosomes	190
9.3	Plasmids		191
	9.3.1	Indigenous plasmids	192
	9.3.2	Exogenous plasmids	193
9.4	Genetic transfer		195
	9.4.1	Capsduction	196
	9.4.2	Conjugation	199
	9.4.3	Transformation	204
9.5	Genetic engineering and prospects for future research		205
	9.5.1	Significance	205
	9.5.2	Current cloning capabilities	205
	9.5.3	Prospectus	209

10 Evolutionary roots of anoxygenic photosynthetic energy conversion: HOWARD GEST

10.1	Introduction		215
10.2	Cyclic photophosphorylation: the central feature of anoxygenic photosynthetic energy conversion		216
10.3	Early bioenergetics; fermentation		217
10.4	Evolution of fermentation mechanisms		219
	10.4.1	Accessory oxidant-dependent fermentations	220
	10.4.2	Use of CO_2 in biosynthesis of accessory oxidants	221
	10.4.3	Evolution of the fumarate reduction system	223
	10.4.4	Accessory oxidant-dependent fermentation in phototrophic bacteria	225

10.5	A model for the origin of anaerobic photophosphorylation	227
	10.5.1 Enslavement of photochemistry by biochemistry	228
10.6	A synopsis of the consequences of invention of anaerobic photophosphorylation	230
10.7	Coda	232

Index   237

Contributing Authors

H. BALTSCHEFFSKY Department of Biochemistry, Arrhenius Laboratory, University of Stockholm, S-106 91 Stockholm, Sweden.

H. H. BEEFTINK Laboratory for Microbiology, University of Amsterdam, Nieuwe Achtergracht 127, 1018 WS Amsterdam, The Netherlands.

M. C. W. EVANS Department of Botany and Microbiology, University College London, Gower Street, London WCIE 6BT, England.

H. VAN GEMERDEN Microbiology Department, Biological Centre, University of Groningen, Kerklaan 30, Haren (Gr.), The Netherlands.

H. GEST Department of Biology, Indiana University, Bloomington, Indiana 47405, U.S.A.

T. A. HANSEN Microbiology Department, Biological Centre, University of Groningen, Kerklaan 30, Haren (Gr.), The Netherlands.

P. HEATHCOTE Department of Botany and Biochemistry, Westfield College, Kidderpore Avenue, London NW3 7GT, England.

B. C. JOHANSSON Department of Biochemistry, Arrhenius Laboratory, University of Stockholm, S-106 91 Stockholm, Sweden.

B. C. MARRS Department of Biochemistry, St. Louis University Medical Center, 1042 South Grand Blvd., St. Louis, Mo 63104, U.S.A.

S. NORDLUND Department of Biochemistry, Arrhenius Laboratory, University of Stockholm, S-106 91 Stockholm, Sweden.

J. OELZE Institut für Biologie II der Universität Freiburg, Schänzlerstr. 1, D-7800 Freiburg, Federal Republic of Germany.

J. G. ORMEROD Botany Department, Oslo University, Blindern, Oslo 3, Norway.

R. SIREVÅG Botany Department, Oslo University, Blindern, Oslo 3, Norway.

Preface

Although it is just one hundred years since the discovery of the anoxygenic phototrophic bacteria, our concept of them seems to have matured during the last five or ten years. This is due partly to our greatly improved knowledge of their energy metabolism but perhaps equally to the vast increases in general understanding of the bacterial cell. Whatever the reasons, the time seemed ripe for a book of this kind.

The book is intended for postgraduate and advanced undergraduate students as well as research workers requiring an orientation on anoxygenic phototrophs. In deciding on chapter headings, emphasis has been placed on aspects of the phototrophic bacteria which make them stand out as a group. Hopefully this will give readers an idea of the very special character of these organisms.

Thanks are due to all the contributors, particularly Mike Evans for timely assistance in a difficult situation, to Bob Campbell and Bridget Cook and the publishers, to Ann-Marie Smit for ever willing help with the typing and finally, to my wife Eli for her forbearance.

University of Oslo

John Ormerod
June 1983

Chapter 1. Introduction

J. G. ORMEROD

About twenty kilometers south of the town of Oslo in Norway, there is a beautiful clear water lake called Polden. When samples are taken from the depths of this lake, it is uncanny to find the water anaerobic, gassy and evil-smelling. To the uninitiated, the circumstance verges on the bizarre when a carefully taken sample from the upper part of this anaerobic zone turns out to be bright green in colour.

The colour is due to a massive population of phototrophic green sulphur bacteria. Some lakes of this type have similar blooms of purple phototrophic bacteria. These obscure bacteria contribute significantly to the primary production and sulphur cycle of the stratified lake ecosystems that support them.

The green, and particularly the purple bacteria have been the object of much research in recent years. The reason for this is that although their photosynthesis is in principle like that of the green plants, it is simpler. Also, bacteria are generally more amenable to experimental investigation than plants.

Although the phototrophic bacteria were first recognized as such about a hundred years ago, research on them was in a confused state for fifty years. The confusion was partly due to the fact that these bacteria do not produce oxygen as the green plants do.

An explanation for this was suggested by the great microbiologist C B van Niel (1930, 1935). He reasoned that CO_2 in the bacterial photosynthesis was being reduced with reductants other than water, e.g. H_2S:

$$CO_2 + 2H_2S \xrightarrow{light} (CH_2O) + H_2O + 2S$$

(CH_2O is a rough expression for cell material.)

The clue to this idea may have been the sulphur, which accumulates in the form of highly refractile globules. Van Niel did careful measurements of the components of the above reaction and confirmed its stoichiometry with pure cultures. But the theory did not stop there: it provided the basis for a unifying concept of photosynthesis because the above equation applies also to the green plant process if the sulphur is replaced by oxygen:

$$CO_2 + 2H_2O \xrightarrow{light} (CH_2O) + H_2O + O_2$$

Thus, van Niel's general equation is:

$$CO_2 + 2H_2A \xrightarrow{light} (CH_2O) + H_2O + 2A$$

This idea was developed at a time when the belief was still fairly new, that most biological oxidations are really stepwise dehydrogenations. In accordance with this, van Niel broke down his general equation so that it included a light-driven dehydrogenation of water:

$$4H_2O \xrightarrow{light} 4H + 4OH$$
$$CO_2 + 4H \longrightarrow CH_2O + H_2O$$
$$\underline{4OH + 2H_2A \longrightarrow 4H_2O + 2A}$$
$$CO_2 + 2H_2A \longrightarrow CH_2O + H_2O + 2A$$

The import of this hypothesis is profound because it implies that light energy creates reducing and oxidizing components (denoted H and OH above) that drive the rest of the process.

Metabolism of phototrophic bacteria

We know today that van Niel's 'components' are charged molecules of quite high molecular weight. These are disposed in small, highly organized assemblages situated in lipid bilayer membranes. These assemblages were first purified by Reed and Clayton (1968) and are called reaction centres. They are present in large numbers in all phototrophic organisms.

Although chlorophyll molecules lie at the very centre of the photosynthetic process, the actual number of them in each reaction centre is quite small, and most of the light utilized by phototrophic organisms is absorbed by supplementary arrays of chlorophyll and other coloured molecules outside the reaction centres. These arrays are collectively known as the light-harvesting pigments and they are arranged in such a manner as to allow efficient transfer of the absorbed light energy to the reaction centres.

The absorption of light results in a charge separation associated with a reaction centre. An enormous amount of research has gone into clarifying the details of this important process.

Equally important from a biochemical point of view is the question of how this charge separation actually drives the chemistry of cell growth. The energy requirements of biosynthesis are in their simplest form a supply of ATP and reducing power. Cells also need to maintain a suitable proton motive force over the cytoplasmic membrane (Konings & Veldkamp, 1980). All of these requirements are met by the operation of a membrane-bound vectorial electron transport system, with accompanying ATP forming machinery, driven by the photochemically-induced charge separation. For many research workers the application of the chemiosmotic hypothesis of P Mitchell (1961) has cleared the way to a better understanding of the events involved in this vitally important process.

One of the essential requirements for growth is a suitable carbon source. Probably all the anoxygenic phototrophic bacteria can use organic carbon sources for growth. Likewise, the great majority, and possibly all, appear to be able to grow with CO_2 as the only carbon source. Such anabolic diversity is unusual even in bacteria, and may be due to the fact that the phototrophic bacteria can use light as their energy source. This ability allows extremely efficient conversion of organic compounds into bacterial cell material, and there is a wealth of assimilatory mechanisms, particularly in the purple bacteria.

Many of the anoxygenic phototrophic bacteria have been shown to be dinitrogen fixers. Intracellular conditions in these organisms would be expected to favour this strictly anaerobic process, which consumes much energy and reducing power. Dinitrogen fixation and assimilation of ammonia have been studied in only a few species, however, and there may exist great variation in mechanisms within the various groups of phototrophic bacteria.

On the other hand there is little to suggest that the phototrophic bacteria are any different from *Escherichia coli* with respect to macromolecular synthesis (nucleic acids, proteins). However, the inconstant and potentially harmful nature of the energy source (light) means that the biosynthesis of photosynthetic membranes and pigments must be closely regulated. This regulation is complicated because it involves not only light intensity and quality, but also temperature, growth rate and in some cases, oxygen.

In recent years techniques of genetic recombination have been developed in purple bacteria involving the use of plasmids and phage vectors. Interest so far has centred upon photosynthetic pigment synthesis and dinitrogen fixation. This field holds great promise for future investigations of all aspects of the phototrophic bacteria.

Diversity of phototrophic bacteria

Up to the publication of the eighth edition of Bergey's (1974) Manual, the anoxygenic phototrophic bacteria were classified under the sub-order Rhodobacteriineae, which was divided into three families: the Thiorhodaceae, or purple sulphur bacteria; the Athiorhodaceae, or purple non-sulphur bacteria; and the Chlorobacteriaceae, or green sulphur bacteria.

The present classification places all the anoxygenic phototrophic bacteria in the order Rhodospirillales, which is divided into four families: the Rhodospirillaceae, or purple non-sulphur bacteria; the Chromatiaceae, or purple sulphur bacteria; the Chlorobiaceae, or green sulphur bacteria; and the Chloroflexaceae or gliding green bacteria.

Recent data on base sequences of 16S ribosomal RNA (Fox *et al.*, 1980) indicate that from a phylogenetic point of view there may be serious faults in the above classification. First, the purple and green bacteria and the two types of green bacteria themselves appear to be much more distantly related to each other than was previously thought.

Second, there are actually *three* groups of purple bacteria: two of the non-sulphur type and one sulphur. Here, the RNA evidence (which is supported by other evidence) shows that the division in the non-sulphur purple bacteria cuts right across the traditional generic boundaries, which are based on morphological characteristics.

With this new division in the genera in mind, it may be possible to uncover previously unsuspected physiological and genetic differences between them. However, until this has been done, and for the purpose of the present discussion, the purple non-sulphur bacteria will be treated as a single group with regard to physiological properties.

Two types of pigment are characteristic of phototrophic bacteria: the chlorophylls and carotenoids. The types of these pigments and, in the case of the chlorophylls, localization in the cell, provide the best means of distinguishing between the purple and the green bacteria.

With a few known exceptions, all the purple bacteria have bacteriochlorophyll *a* as their sole chlorophyll (see Fig. 1.1 for structure). The exceptions have bacteriochlorophyll *b* as their sole chlorophyll. In purple bacteria all of the chlorophyll is located in unit membrane structures that are invaginations of the cytoplasmic membrane. The chlorophyll functions both in light harvesting and in the reaction centres.

All known green bacteria possess two types of chlorophyll. Bacteriochlorophyll *a* functions in the reaction centres of the cytoplasmic membrane and is also present as a water soluble complex with protein, associated with the cytoplasmic membrane. Bacteriochlorophylls *c, d* or *e* (previously known as chlorobium chlorophylls) are located in cigar-shaped vesicles called chlorosomes just inside the cytoplasmic membrane. These vesicles are bounded by non-unit membranes and serve a light harvesting function.

Since the bacteriochlorophylls have characteristic absorption spectra and are easily extractable with organic solvents, e.g. methanol, it is a simple matter to determine whether an organism is a purple or a green bacterium by examining the absorption spectrum of its methanol extract.

The phototrophic bacteria contain a great variety of carotenoid pigments. They function as accessory light harvesters and also serve to protect the

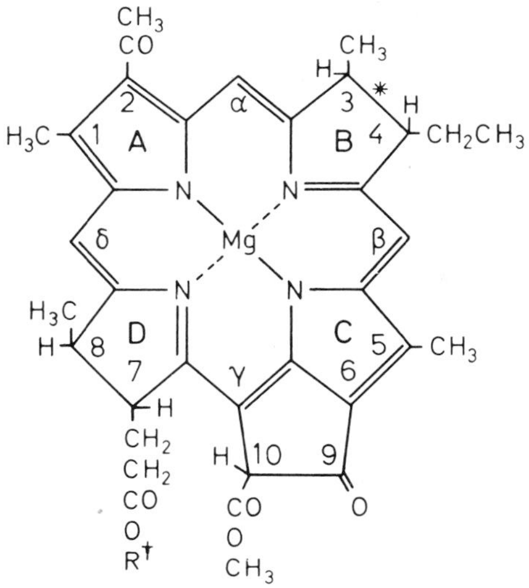

Fig. 1.1 Structure of bacteriochlorophylls. The structure shown is that of bacteriochlorophyll *a*. The other bacteriochlorophylls differ from bacteriochlorophyll *a* as follows:

	Carbon atom					
	C_2	C_3	C_4	C_5	C_{10}	δC
Bacteriochlorophyll						
b			$=CHCH_3$			
c	$-CHOHCH_3$	$-CHO$	alkyl	ethyl	$-H$	$-CH_3$
d	$-CHOHCH_3$	$-CHO$	alkyl	ethyl	$-H$	
e	$-CHOHCH_3$	$-CHO$	alkyl	ethyl	$-H$	$-CH_3$

* double bond between C_3 and C_4 in bacteriochlorophyll *c*, *d*, and *e*.
† R = phytyl in bacteriochlorophyll *a* and *b*; farnesyl in bacteriochlorophyll *c*, *d* and *e*.

photosynthetic apparatus against photo-oxidation. Some of the carotenoids are characteristic of families, genera and even species of phototrophic bacteria, and together with the chlorophylls they are largely responsible for the various colours of organisms within the group. Thus, some 'purple' bacteria may be brown, or even green. Likewise, green bacteria may be brown or even orange, as in the case of *Chloroflexus* growing in bright light.

A general character of the purple sulphur bacteria is growth on sulphide and storage of elementary sulphur intracellularly. They can also grow on simple organic compounds anaerobically in the light. While many, perhaps all, of the purple non-sulphur bacteria can grow on sulphide (mostly at low concentrations) they all grow best on simple organic substrates. Many of them are facultative anaerobes and can grow on such substrates in the dark under aerobic conditions. Provided the aeration is good, the formation of the photosynthetic pigments is repressed, resulting in whitish cells.

On the other hand, the green sulphur bacteria are killed by exposure to air. They grow only in the light and oxidize H_2S to H_2SO_4, excreting elementary sulphur as an intermediate into the medium. They can assimilate a few simple organic compounds if CO_2 and H_2S are also present.

In contrast, the gliding, green bacteria thrive on organic media both anaerobically in the light and aerobically in the dark. Thus, physiologically they appear similar to the purple non-sulphur bacteria.

At this point it is appropriate to mention the recently discovered ability of certain cyanobacteria to grow in a manner quite analogous to that of the purple and green sulphur bacteria, i.e. carrying out an anoxygenic photosynthesis with accessory electron donors like H_2S and thiosulphate (Cohen et al., 1975; Utkilen, 1976). This apparently atavistic property suggests a common ancestry for all phototrophic bacteria, anoxygenic and oxygenic alike.

Cell motility might be regarded as an important attribute of aquatic phototrophs. Accordingly, most of the purple bacteria are motile, with polar flagella, and the Chloroflexaceae can move by a gliding mechanism. On the other hand, the green sulphur bacteria are immotile. Gas vacuoles are present in representatives of both green and purple bacteria. This may explain how the non-flagellated types are able to maintain themselves at a favourable level in a water column.

Table 1.1

Bacteria	Genera
Purple non-sulphur Family Rhodospirillaceae	*Rhodospirillum* *Rhodopseudomonas* *Rhodomicrobium* *Rhodocyclus*
Purple sulphur Family Chromatiaceae	*Chromatium* *Thiocystis* *Thiosarcina* *Thiospirillum* *Thiocapsa* *Lamprocystis* *Thiodictyon* *Thiopedia* *Amoebobacter* *Ectothiorhodospira*
Green sulphur Family Chlorobiaceae	*Chlorobium* *Prosthecochloris* *Pelodictyon* *Clathrochloris*
Green gliding Family Chloroflexaceae	*Chloroflexus* *Chloronema* *Oscillochloris*

The above table is included for the purpose of reference. It is based primarily on information in the eighth edition of Bergey's (1974) Manual.

REFERENCES

BERGEY'S (1974) *Manual of Determinative Bacteriology* (Eds. Buchanan R. E. & Gibbons N. E.). 8e. pp. 1268. Williams & Wilkins Co., Baltimore.

COHEN Y., PADAN, E. & SHILO, M. (1975) Facultative bacteria-like photosynthesis in the blue-green alga *Oscillatoria limnetica*. *J. Bact.*, **123**, 855–61.

FOX G. E., STACKEBRANDT E., HESPELL R. B., GIBSON J., MANILOFF J., DYER T. A., WOLFE R. S., BALCH W. E., TANNER R. S., MAGRUM L. J., ZABLEN L. B., BLAKEMORE R., GUPTA R., BONEN L., LEWIS B. J., STAHL D. A., LUEHRSEN K. R., CHEN K. N. & WOESE C. R. (1980) The phylogeny of the prokaryotes. *Science*, **209**, 457–63.

KONINGS W. N. & VELDKAMP H. (1980) Phenotypic responses to environmental change. In *Contemporary Microbial Ecology* (Eds. Ellwood D. C., Hedger J. M., Latham M. J., Lynch J. M. & Slater J. H.) pp. 161–91. Academic Press, London & New York.

MITCHELL P. (1961) Coupling of phosphorylation to electron and hydrogen transfer by a chemiosmotic type of mechanism. *Nature, Lond.* **191**, 144–8.

VAN NIEL C. B. (1930) Photosynthesis of bacteria. In *Contributions to Marine Biology.*, pp. 161–9. Stanford University Press.

VAN NIEL C. B. (1935) Photosynthesis of bacteria. *Cold Spring Harb. Symp. quant. Biol.* **3**, 138–50.

REED D. W. & CLAYTON R. K. (1968) Isolation of a reaction centre fraction from *Rhodopseudomonas sphaeroides*. *Biochem. Biophys. Res. Commun.*, **30**, 471–5.

UTKILEN H. C. (1976) Thiosulphate as electron donor in the blue green alga *Anacystis nidulans*. *J. gen. Microbiol.*, **95**, 177–80.

Chapter 2. Structure of Phototrophic Bacteria; Development of the Photosynthetic Apparatus

J. OELZE

2.1 INTRODUCTION

The distinctive character of the energy metabolism of most phototrophic bacteria is associated with a special type of intracellular membrane which contains the photosynthetic apparatus. Apart from this, the phototrophic bacteria are structurally like other Gram-negative bacteria and comprise a range of morphological types.

To distinguish the internal, photosynthetic membranes from the peripheral cytoplasmic membrane they have been named intracytoplasmic membranes (Holt & Marr, 1965a; b). Upon cell homogenization the intracytoplasmic membrane system usually breaks down into single vesicles which, because of the presence of the brightly coloured pigments of the photosynthetic apparatus, are referred to as chromatophores (Schachman et al., 1952). In the following paragraphs the term chromatophore is employed to denote isolated membrane preparations derived from the photosynthetically active membrane system of whole cells.

Some members of the Rhodospirillales are facultative phototrophs which perform a chemotrophic energy metabolism under aerobic conditions. The respiratory system responsible for this is located predominantly in the cytoplasmic membrane. Thus, typical cells of the facultatively phototrophic bacteria do not produce photosynthetically active intracytoplasmic membranes when growing under conditions of high aeration. The synthesis of bacteriochlorophyll* is inhibited by oxygen (see Section 2.4.1). Consequently, Bchl and the photosynthetic apparatus are formed only when cells are transferred from high to low oxygen tensions or even to anaerobiosis. Of course, the photosynthetic apparatus functions only when the cultures are illuminated. This simple manipulation of culture conditions provides a convenient basis for studying the processes involved in the reversible

---

* Abbreviations used in this chapter: Bchl, bacteriochlorophyll; LH, light-harvesting; RC, reaction centre; δ-Ala, δ-aminolevulinic acid.

adaptation from a chemotrophic to a phototrophic energy metabolism, including the development of the bacterial photosynthetic apparatus and its membrane as a structural counterpart.

Once formed, the photosynthetic apparatus is not necessarily stable but is subject to qualitative and quantitative changes depending on such environmental conditions as may influence the cellular energy metabolism. This applies to facultatively as well as obligatorily phototrophic species.

It is the object of this chapter to present basic aspects of the current knowledge on the structure, composition and development of the photosynthetically active membranes of phototrophic bacteria.

2.2 FINE STRUCTURE OF CELLS AND MEMBRANES

Electron micrographs of thin-sections of most species of purple phototrophic bacteria reveal the presence of intracytoplasmic membranes (Oelze & Drews, 1972). A few members, like *Rhodospirillum tenue* or *Rhodopseudomonas gelatinosa*, generally show no intracytoplasmic membranes, although occasionally single irregularly shaped membrane invaginations occur (de Boer, 1969).

Furthermore, none of the Chlorobiineae produces typical intracytoplasmic membranes. Instead, these bacteria are characterized by special structures, the chlorosomes, which underlie the cytoplasmic membrane (Cohen-Bazire, 1963; Staehelin *et al.*, 1978). Chlorosomes, as is described below (Section 2.3), house the LH pigments of the photosynthetic apparatus, but they are not bounded by a unit membrane (Cohen-Bazire, 1963). The number of chlorosomes in green bacteria and the cellular amount of intracytoplasmic membranes in purple bacteria vary inversely with light flux (reviews by Lascelles, 1968 and Oelze, 1981).

Most members of the Rhodospirillaceae and Chromatiaceae contain vesicular intracytoplasmic membranes (Fig. 2.1) (reviewed by Remsen, 1978). From studies on intracytoplasmic membranes carefully liberated from cells it has been inferred that the vesicular structures, visible in thin sections, are part of a regularly bulged tubular intracytoplasmic membrane reticulum (Tuttle & Gest, 1959; Holt & Marr, 1965a). Nevertheless, the possibility does exist, that some of the bulges separate from the reticulum and become single closed vesicles in the cytoplasm.

Vesicular intracytoplasmic membranes are present in many species of purple bacteria, including *Rs. rubrum*, *Rp. capsulata*, *Rp. sphaeroides* and various *Chromatium* species. Unbulged tubular intracytoplasmic membranes arranged in bundles have been observed in *Thiocapsa pfennigii* (Eimhjellen *et al.*, 1967).

Another relatively large group of the phototrophic bacteria exhibits stacked lamellar intracytoplasmic membranes (Fig. 2.1). This group is represented by

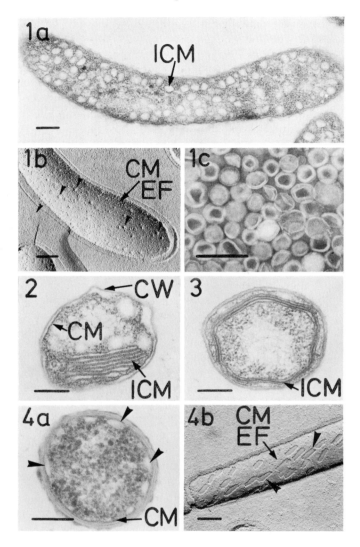

Fig. 2.1 Electron micrographs of phototrophic bacteria. Bars represent 200 nm.
1. *Rhodospirillum rubrum* grown phototrophically under anaerobic conditions.
(a) Longitudinal section of a cell with numerous intracytoplasmic membrane (ICM) vesicles.
(b) Freeze-fracture preparation exposing the exoplasmic fracture face (EF) of the cytoplasmic membrane (CM) with numerous invaginations (arrow heads) to form intracytoplasmic membranes (Golecki & Oelze, 1975). (c) Negative staining preparation of isolated intracytoplasmic membrane vesicles.
2. Ultrathin section of *Rhodopseudomonas viridis* exposing a stack of intracytoplasmic membranes. (CW = cell wall)
3. Cross section of *Rhodomicrobium vannielli*.

Rs. fulvum and *Rs. molischianum*, *Ectothiorhodospira mobilis*, *Rhodomicrobium* and *Rp. viridis*, *Rp. acidophila* and *Rp. palustris*.

Occasional connections can be observed between cytoplasmic and intracytoplasmic membranes in electron micrographs of thin-sectioned cells (reviewed by Remsen, 1978). Freeze-fracture electron microscopy of *Rs. rubrum* reveals indentations on the convex (plasmic) fracture face and protuberances on the concave (exoplasmic) fracture face of the cytoplasmic membrane. These have been interpreted to represent the connections between cytoplasmic and intracytoplasmic membranes (Golecki & Oelze, 1975). Because of these connections the periplasmic space is continuous with the interior of the intracytoplasmic membrane vesicles.

Electron micrographs of thin-sections of cells and membrane preparations from *Rs. rubrum*, *Rp. capsulata* and *Rp. sphaeroides* show the presence of knob-like particles (Löw & Afzelius, 1964; Lampe *et al.*, 1972; Reed & Raveed, 1972) which are thought to represent the head pieces of coupling factor ATPase (Reed & Raveed, 1972). More recently, freeze-fracture electron microscopy has been employed to study the architecture of membranes on the basis of distribution and size of intramembrane particles, revealing significant differences in the specific architecture of the membranes of *Rs. rubrum*, *Rp. capsulata*, *Rp. sphaeroides* and *Rs. tenue* (Golecki *et al.*, 1979; Golecki & Oelze, 1980). However, in principle, the following results were obtained with all of these species:

1. photosynthetic membranes exhibited higher numbers of particles per unit area than respiratory membranes;
2. except for *Rs. tenue*, higher numbers of particles were expressed on the plasmic than on the exoplasmic leaflet of the membranes.

Unfortunately, a correlation between specific particles and functions is not available at present. A determination of this correlation, however, will be one of the future goals in the study of membrane structure.

2.3 COMPOSITION AND STRUCTURE OF CHROMATOPHORE VESICLES

Chromatophores isolated from various purple bacteria have been subjected to chemical and physical analyses and a fairly detailed picture of their structural and functional properties has been developed. But it should be kept in mind

---

4. *Chlorobium limicola*. (a) Thin section showing several chlorosomes (Staehelin *et al.*, 1978) appressed to the cytoplasmic membrane. (b) Freeze-fractured cell presenting the exoplasmic fracture face (EF) of the cytoplasmic membrane with base plates (arrow heads) of associated chlorosomes.
(Electron micrographs courtesy of Dr. J. R. Golecki)

Table 2.1 Physical properties and gross chemical composition of chromatophore vesicles from selected representatives of the phototrophic bacteria (Lit. compiled by Oelze, 1981).

Species	Diameter (nm)	Buoyant densities (gcm^{-3})	Weight (μg × 10^{-11})	Protein*	Total* lipid	Phospho-* lipid	Bacterio-* chlorophyll	Carotenoids
Rs. rubrum	70–90	1.165	7.8	55	n.d.†	15.6	3.2	n.d.†
Rp. sphaeroides	60	1.16	6.3	63	30	n.d.†	5.9	1.9
C. vinosum	60	1.17	2.1	27	67	n.d.†	8.3	1.3

* Values are presented as percent of dry weight.
† n.d. = not determined.

that membranes of a living system are not fixed structures; they may adapt dynamically to intracellular as well as extracellular changes (see Section 2.4). The data to be presented here were obtained with membrane preparations isolated for the most part from cells in the late phase of exponential growth.

Table 2.1 shows buoyant densities and ratios of protein to lipid which are largely comparable with those of membranes of other biological origins. But, more detailed analyses indicate specific properties, e.g. Bchl and carotenoid content, which reflect the special photosynthetic function. Carotenoids are the pigments largely responsible for the colours of purple and brown bacteria and, consequently, of chromatophore preparations derived from them. Carotenoids function either in the light absorption process or by channelling superfluous energy out of the system, thereby preventing harmful photo-oxidations (Cogdell *et al.*, 1976; Amesz, 1978). At least 75 different carotenoids have been identified from phototrophic bacteria, but a detailed description of these pigments is outside the scope of this chapter. Nevertheless, it should be mentioned that carotenoids of phototrophic bacteria have several unique chemical properties. Schmidt (1978) compiled the following list of properties characteristic of the carotenoids of phototrophic bacteria:

(i) they are mostly aliphatic;
(ii) many contain tertiary hydroxyl and methoxyl groups;
(iii) double bonds in the C-3,4 position are frequent;
(iv) oxogroups are found on C2 or C4 of the polyene chain;
(v) aldehyde groups occur in the C-20 position, and
(vi) the cyclic carotenoids commonly have aromatic rings.

The second group of isoprenoid lipids of phototrophic bacteria, the quinones, function as membrane-bound carriers in electron transport. Both ubiquinones and menaquinones have been identified in phototrophic bacteria (Parson, 1978).

The predominant phospholipids identified to date comprise phosphatidyl glycerol, phosphatidyl ethanolamine and phosphatidyl choline. The Chlorobiineae are unique among the anoxygenic phototrophic bacteria in that they contain glycolipids in addition to phospholipids. Analysis of phototrophic bacterial lipids shows the presence of myristic (14:0), palmitic (16:0) and minor amounts of stearic (18:0) acid. In addition the unsaturated palmitoleic (16:1) and *cis* vaccenic (18:1) acids have been identified. Species like *Rs. rubrum*, *Rp. sphaeroides*, *Rp. capsulata* and *Chromatium vinosum*, which produce intracytoplasmic membranes, exhibit a remarkably high proportion, up to 90%, of unsaturated fatty acids. On the other hand, the ratio of unsaturated to saturated fatty acids is about one in species like *Rs. tenue* and members of the Chlorobiineae, which form few, if any, intracytoplasmic membranes. A detailed review of the lipids of phototrophic bacteria has been given by Kenyon (1978).

Membrane proteins are generally assumed to be integrated into functional units such as the various transport or metabolic systems. This is also true for phototrophic bacteria. The photosynthetic apparatus, however, contains additional pigment protein complexes which transform light energy into electronic energy and, moreover, into energy stored in the primary photochemical charge separation. These complexes contain Bchl, and more than 60% of the polypeptides determined after sodium dodecyl sulphate polyacrylamide gel electrophoresis of solubilized membranes of purple bacteria can be attributed to Bchl *a* functional complexes (reviewed by Oelze, 1981). These comprise photochemical RCs and LH antenna Bchl complexes.

Light-harvesting complexes occur in the phototrophic bacteria in different forms. While *Rs. rubrum* exhibits only one type, with a Bchl *a* absorption peak at about 875–880 nm (designated B875), most of the other Rhodospirilliineae produce in addition, an accessory LH complex. In most *Rhodopseudomonas* species this accessory Bchl *a* complex absorbs light in constant proportions at 800 and 850 nm and is designated B800–850. In contrast, peak heights may vary in LH complexes from *Chromatium vinosum* and *Rp. acidophila* strain 7050 and in these species a third absorption band may occur at 820–830 nm (Mechler & Oelze, 1978b; Cogdell *et al.*, 1979). In the Chlorobiineae the chlorosomes house accessory LH pigment (Olson *et al.*, 1977). In *Chlorobium limicola* and *Chloroflexus aurantiacus* this is Bchl *c*, while the cytoplasmic membrane of these organisms contains Bchl *a* associated with RC and LH units (Olson *et al.*, 1977; Pierson & Castenholz, 1978; Schmidt, 1980). The size of the photosynthetic unit in green sulphur bacteria is considerably greater than in purple bacteria (Fowler *et al.*, 1971). Less is known about the photosynthetic unit in *Chloroflexus*. The LH-Bchl *a* protein complex of green sulphur bacteria is the only such complex to have been crystallized. It has been thoroughly characterized by X-ray diffraction (Fenna & Matthews, 1975) and its subunit is described as a string bag of protein with seven Bchl *a* molecules inside (see Section 3.3.1 for more details). Future investigation will show whether this type of structure is characteristic of LH-chlorophyll complexes in general.

To date, no pure LH complexes have been obtained from the few organisms which produce Bchl *b*. However, in *Rp. viridis* the absorption band at about 1015 nm represents the antenna Bchl *b* (Fig. 2.2) (Thornber *et al.*, 1978). Bchl *b* contained in RC preparations from *Rp. viridis* absorbs at 830 and 960 nm (Fig. 2.2) (Clayton & Clayton, 1978).

Data on the chemical composition of photochemical RC's and LH Bchl complexes isolated from *Rp. sphaeroides* are compiled in Table 2.2. The absorption properties and polypeptide composition of the RC's from *Chromatium vinosum*, *Rp. capsulata* and *Rs. rubrum* are very similar to those of *Rp. sphaeroides* (Feher & Okamura, 1978; Mechler & Oelze, 1978a).

The arrangement in the membrane of the constituents of membrane-bound reaction systems is of great importance for their function. Investigations

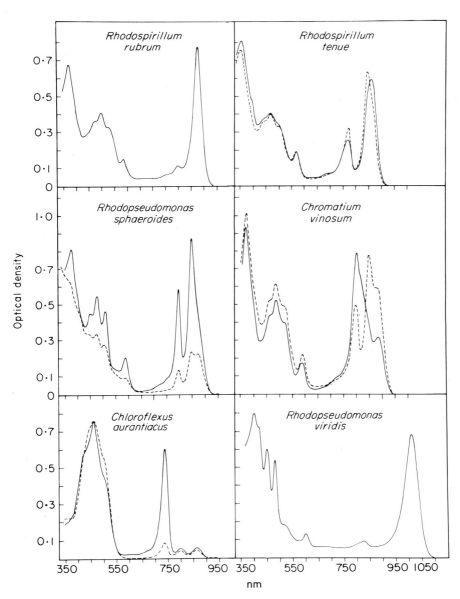

Fig. 2.2 Absorption spectra of membrane preparations from different members of the phototrophic bacteria. Solid lines: representative of cultures grown at low light fluxes; broken lines: representative of cultures grown at high light fluxes. For *Rhodospirillum rubrum* and *Rhodopseudomonas viridis* no spectral changes in the near infrared have been reported to date. *Rhodospirillum rubrum*, *Rhodospirillum tenue*, *Rhodopseudomonas sphaeroides*, and *Chromatium vinosum* form Bchl *a*; *Chloroflexus aurantiacus* synthesizes Bchl *a* and *c* (peak at 740 nm; the ratio of Bchl *c* to *a* was 5 under low and 1.2 under high light flux); *Rhodopseudomonas viridis* produces Bchl *b*.

Table 2.2 Composition of functional units isolated from membranes of *Rhodopseudomonas sphaeroides*.

Species	Units	Mol. wt.	Mol. wt. of polypeptides (after SDS-PAGE)*	Constituents (molar ratio)	References
Rhodopseudomonas sphaeroides	Photochemical reaction centre	92 000	28 000 24 000 21 000	4 Bacteriochlorophyll 2 Bacteriophaeophytin Fe, Ubiquinone	Feher & Okamura, (1978)
	Light-harvesting B875-complex		12 000 8000	2 Bacteriochlorophyll 2 Carotenoid	Broglie *et al.*, (1980)
	Light-harvesting B850-800-complex	22 000	10 000 8000	3 Bacteriochlorophyll 1 Carotenoid	Sauer & Austin, (1978)

* Sodium dodecyl sulphate polyacrylamide gel electrophoresis.

on the topography of chromatophores from *Rs. rubrum*, *Rp. sphaeroides* and *Rp. capsulata* have provided the following information: homogenization of cells with a French pressure cell or by osmotic lysis liberates closed vesicles from the originally vesicular intracytoplasmic membrane system, practically without reversion of the original membrane orientation.

Thus, the outer surface of the chromatophores represents the cytoplasmic face of the intracytoplasmic membrane system, while the inner face of the chromatophores represents the periplasmic face. From studies employing other organisms, it is known that most of the functional units in a membrane are exposed at the cytoplasmic surface (Salton & Owen, 1976). This is true also for chromatophores. In particular ATPase (F_1), succinic-, lactate-, and NADH-dehydrogenases are exposed at the outer surface of chromatophores, i.e., the cytoplasmic face of intracytoplasmic membranes. Furthermore, several investigations point to the accessibility of photochemical RC's at the chromatophore outer surface. Detailed analyses, however, revealed the exposure of the heavy rather than of the two other lighter polypeptides (Table 2.2). More recent results suggest that the heavy polypeptide spans the membrane while the two lighter ones are embedded in the hydrophobic interior of the membrane (Bachofen, 1979).

Amino acid analyses of LH protein complexes show a rather low polarity. Although this implies that the protein is embedded in the lipophilic interior of the membrane, at least a partial exposure of this unit at the outer face is strongly suggested by topographical studies.

Little information is available on the exposure of functional units at the inner (periplasmic) face of chromatophores. The only firmly established result points to a localization of cytochrome c_2 at the inner face of chromatophores from *Rp. capsulata* and *Rp. sphaeroides*. This is of great importance for the proper functioning of the vectorial photochemical electron transport and proton translocation across the membrane (Prince *et al.*, 1975).

According to current hypotheses on membrane structure, proteins penetrate the phospholipid bilayer to different extents. Lateral movement of the proteins and lipids may be facilitated through the fluidity of the lipid bilayer. But, investigations by Birrell *et al.* (1978) and Fraley *et al.* (1978a) indicate that the mobility of membrane lipids is greatly reduced by the presence of proteins from the photosynthetic apparatus. The reader is referred to Chapter 4 and to reviews by Drews and Oelze (1980) and Oelze (1981) for more detailed treatments.

2.4 DEVELOPMENT OF THE PHOTOSYNTHETIC APPARATUS

In a culture of a phototrophic bacterium growing in the light, the amount and composition of the photosynthetic apparatus are very closely regulated.

Under steady-state conditions, this regulation results in the establishment of a balance between light energy absorbed and energy consumed in the reactions of maintenance and growth.

Four major factors are known to be involved in this regulation:

(1) light flux
(2) temperature
(3) growth rate, as influenced by nutrition
(4) partial pressure of oxygen (in facultative phototrophs).

As mentioned in Section 2.1, the biosynthesis of the photosynthetic apparatus has been studied almost exclusively with facultatively phototrophic bacteria. These form the photosynthetic apparatus when transferred from chemotrophic conditions with high aeration to anaerobic conditions in the light or conditions of low aeration in the dark. Apparently, regulatory devices which control the onset of the synthesis of the entire photosynthetic apparatus require anaerobiosis or at least lowered oxygen partial pressures. In Section 2.4.1 it is shown that this results from the sensitivity toward oxygen of Bchl synthesis. This implies that the production and incorporation into existing membranes of the different Bchl complexes is the process necessary to convert a membrane with respiratory activities into a photosynthetically active one.

In fact, in testing the minimum requirements necessary to convert Bchl-depleted membranes *in vitro* into photosynthetically active membranes, Garcia and Jones and colleagues demonstrated that incorporation of photochemical RC's was sufficient to reconstitute photochemical electron transport (Garcia *et al.*, 1975; Hunter & Jones, 1979a; b). This supports the findings reviewed by Jones (1976) that electron transport components of the respiratory system may participate in photochemical electron flow. Because of this central role of Bchl in the adaptation to phototrophy, a brief description of the regulation of Bchl synthesis is presented before describing the development of photosynthetic membranes.

2.4.1 Regulation of bacteriochlorophyll synthesis

In 1957, Cohen-Bazire *et al.* reported that transfer of phototrophically grown cells of non-sulphur purple bacteria from dim to bright light inhibited Bchl synthesis. Moreover, aeration of anaerobically grown cultures also inhibited Bchl formation. Later, this inhibition of Bchl production by increased light fluxes or oxygen tensions was shown to be general within the entire group of the phototrophic bacteria although some species were more sensitive than others. For example, *Rp. capsulata*, *Rp. palustris* and *Rs. tenue* still form small but measurable quantities of Bchl when growing in a medium saturated with air, while *Rs. rubrum* does not (reviewed by Oelze, 1981). It is noteworthy, that, in those species which form two types of LH units, i.e., B875 and B800–850 (see

Section 2.3), the formation of the latter type is usually more sensitive to inhibition by increased light or oxygen tension than that of B875. On the other hand, data available so far point to a largely coordinated synthesis of B875 and photochemical RC Bchl's (Sistrom, 1978).

In the period since the early investigations on the *in vivo* regulation of Bchl formation, various hypotheses have been developed to explain the mechanism of regulation. In essence, these claim the involvement of either the reduction-oxidation state of the electron transport system or of the state of cellular energy metabolism (reviewed by Drews & Oelze, 1980). For example, it has been proposed that Bchl is not synthesized if either the oxidation state of the presumed carrier or the rate of ATP regeneration exceed a certain level. Both conditions may become effective after transfer of cells from low to high light fluxes. While the inhibitory effect of oxygen favours the hypothesis of the regulatory properties of the oxidation state of an effector molecule, this does not exclude parallel regulatory mechanisms. However, the balance of evidence for regulation of Bchl synthesis by ATP-regeneration rates or energy charge is at present negative.

Clearly, new experimental approaches must be designed before conclusive evidence can be obtained on the regulation of Bchl synthesis. One further step in this direction is the investigation of the regulation of enzymes of the Bchl biosynthetic pathway. In *Rp. sphaeroides*, Bchl synthesis is both initiated and rate-limited by the activity of δ-Ala synthase which catalyses the formation of δ-Ala from glycine and succinyl-CoA (Lascelles, 1978). Regulation of δ-Ala synthase activity has been studied extensively and it has become evident that the purified enzyme is inhibited by compounds of obvious physiological significance like haem, protoporphyrin, Mg-protoporphyrin, and ATP. Moreover, δ-Ala synthase can be present in either a more or a less active form. Activation may be regulated by the concentration of trisulphides which, in turn, may be linked to electron transport (Sandy *et al.*, 1975; Oyama & Tuboi, 1979).

In spite of these results, it should be kept in mind that δ-Ala is required for the synthesis not only of Mg-tetrapyrroles, i.e., the chlorophylls, but also of Fe and Co-tetrapyrroles. Thus, there must be special mechanisms regulating either the Mg or the Fe or Co branches. At the stage succeeding the formation of protoporphyrin IX, the Mg and Fe branches separate through the activity of the respective chelatases which incorporate the proper metal ion into the tetrapyrrole system. Incorporation of Mg appears to be particularly sensitive to oxygen, although practically nothing is known about the regulation of the final steps in the synthesis of Bchl and its associated proteins. However, since both Bchl and the protein moieties are always present together, their formation is probably coordinated.

Besides regulation of enzyme activities, repression and derepression of enzyme synthesis may be involved in the regulation of Bchl synthesis. Indeed, Lascelles (1978) reviewed evidence on the repression by aeration of δ-Ala synthase, Mg protoporphyrin methyl transferase and a third enzyme which incorporates Mg into protoporphyrin.

In summary then, several lines of evidence obtained from studies on whole cells and enzyme reactions point to the participation of redox-reactions in the regulation of Bchl synthesis. This type of regulation enables some of the facultatively phototrophic members to produce the photosynthetic apparatus in the dark under conditions of either low oxygen tension or even of anaerobiosis (Kaplan, 1978; Uffen, 1978). This dark Bchl formation appears to be gratuitous under laboratory conditions but it is no doubt of ecological importance. It is easy to see the advantages to be gained by a bacterium growing phototrophically in daylight, if it can continue to synthesize the photosynthetic apparatus at night and/or under low oxygen tensions.

2.4.2 Formation of reaction centre and light-harvesting bacteriochlorophyll units

The photochemical RC's contain only a minor fraction of the total Bchl. The near infrared absorption bands of whole cells or subcellular fractions are therefore due predominantly to the LH units. A comparison of near infrared absorption spectra of the different phototrophic bacteria reveals a diversity with respect to not only the position but also the relative heights of the absorption bands (Fig. 2.2). This applies both to organisms containing two types of Bchl—which show *per se* different absorption properties—and to species producing only one Bchl type. In the latter case, variation in absorption properties arises from the mode of integration of Bchl within different functional units. (For a more detailed discussion of this see Thornber *et al.*, 1978; Thornber & Barber, 1979).

Two methods are employed at present to quantitate the formation of photochemical RC and LH Bchl *a* units. The first is based on the fact that upon photo-oxidation, the absorption band of RC Bchl *a* at 865–880 nm is bleached.

Although the exact position of the bleachable absorption band of RC Bchl *a* varies depending on the different species discussed in the following paragraphs, these RC's are referred to uniformly as P_{865}. Difference absorption coefficients have been determined which allow quantitative estimation of RC's (Clayton, 1966; Straley *et al.*, 1973). The ratio between total Bchl *a* (which is mostly LH Bchl) and RC Bchl *a* defines the size of the photosynthetic unit (Sistrom, 1978). The amounts of the two Bchl *a* units, i.e., B875 and B800–850 can be found from the peak heights at 800, 850 and 875 nm (Sistrom, 1978).

The second method relies on the quantitation of polypeptides characteristic of RC's and LH units, respectively. Polypeptide patterns can be obtained after electrophoresis of the sodium dodecyl sulphate-solubilized samples on polyacrylamide gels (Takemoto, 1974).

Since the original investigations made by Aagaard & Sistrom (1972), it has been known that, under all sets of conditions that permit Bchl synthesis, *Rs. rubrum* exhibits a photosynthetic unit of fixed composition with about 20–30 mol of total Bchl *a* per mol RC (P_{865}). This RC contains 4 mol of Bchl *a* (see Table 2.2). Thus, in this organism, RC's (P_{865}) and LH Bchl *a* (B875)-units are synthesized in a largely coordinated fashion. This was confirmed by spectral analysis for cells adapting from high to low aeration, although under these conditions, the kinetics of polypeptide formation showed an initial lack of coordination. Nevertheless, after about three hours of adaptation a strict coordination was observed with respect to the formation of polypeptides associated both with RC's and with LH units (Oelze & Pahlke, 1976). This example suggests that the polypeptides of the RC-LH unit need not be present in their characteristic proportions for the unit to be functional.

In *Rp. sphaeroides*, *Rp. capsulata* and *Rp. palustris*, the accessory LH (B800–850) Bchl *a*-complex shows absorption bands which are in constant proportion to each other (reviewed by Drews & Oelze, 1980). Similarly, the absorption bands of the RC (P_{865})-LH (B875) Bchl *a* unit show that its pigment composition also is stable. However, the concentration ratio between RC-LH units and accessory LH-complexes varies with culture conditions. For these variations the following processes appear to be responsible:

1. The adaptive formation of the photosynthetic apparatus is initiated by the preferential production of RC-LH units. After about one hour the enhanced formation of accessory LH complexes can be detected, leading to an increase in the size of the photosynthetic unit (Takemoto, 1974; Nieth & Drews, 1975; Firsow & Drews, 1977).
2. Environmental factors that cause a decrease in the rate of total Bchl synthesis, affect primarily the formation of the accessory LH complexes (Fig. 2.2). These factors include increased oxygen tension or light intensity, and decreased temperature. On the other hand, the rate of B800–850 formation is enhanced along with the rate of total Bchl formation when the oxygen tension or light intensity are decreased or the temperature is increased (Lien *et al.*, 1973; Takemoto & Huang-Kao, 1977; Schumacher & Drews, 1978; Kaiser & Oelze, 1980).
3. These results suggest that in the three above-mentioned species, and only below a certain, critical Bchl level, the regulation of Bchl formation is represented by the production of RC-LH units. At higher Bchl levels regulation occurs through preferential formation of accessory LH-units. The critical value of the cellular Bchl level depends upon the species (reviewed by Oelze, 1981).

In organisms which are predominantly or obligately phototrophic it is difficult or impossible to study the *de novo* synthesis of the photosynthetic apparatus. Studies of adjustment of the photosynthetic apparatus to changes in the environment give useful information in such organisms. For example, *Chromatium vinosum*, with an RC (P_{865})-LH (B875) Bchl *a* unit of fixed composition, exhibits changes in the near infrared spectrum in accordance with light intensity during growth (Fig. 2.2). At high light flux the maximum absorption band is at 850 nm with a secondary peak at 800 nm and a shoulder at 880 nm. At low light flux the maximum absorption is at 800 nm with shoulders at 820, 850 and 880 nm. These differences are due to alterations in the amount and spectral characteristics of an accessory LH Bchl *a* complex. Such complexes isolated from cells grown at high and low light intensities reflect these characteristics (Mechler & Oelze, 1978a; b). In addition, *C. vinosum* adapts to changes in light flux by adjusting the total level of Bchl. However, the ratio of accessory LH Bchl to RC (P_{865})-LH (B875) Bchl *a* units varies at most by a factor of two. A comparable change in the absorption properties was found with *Rp. acidophila* (Cogdell *et al.*, 1979).

In *Rs. tenue*, it is primarily the intermediate Bchl *a* absorption peak—located in this organism at 865 instead of 850 nm—which increases with increasing light energy fluxes (Fig. 2.2) (Oelze & Wakim, unpublished).

In *Rp. viridis*, which produces Bchl *b*, the infrared absorption properties of the photosynthetic apparatus are apparently stable, in spite of changes in environmental conditions (Thornber *et al.*, 1978).

Responses of green phototrophic bacteria toward changes in light energy supply have been investigated with *Chlorobium limicola* and *Chloroflexus aurantiacus*. As might be expected, the ratio of accessory Bchl *c* (contained in chlorosomes) to Bchl *a* (associated with RC's and LH units in the cytoplasmic membrane) varies inversely with the light flux (Fig. 2.2) (see Pierson & Castenholz, 1974, 1978). However, in response to comparable changes in light intensity, the variation in this ratio was different in the two organisms, changing up to six-fold in *Chloroflexus* but only two-fold in *Chlorobium* (Broch-Due *et al.*, 1978; Pierson & Castenholz, 1978). This is paralleled, at least in *Chlorobium*, by variations in the number, size and density of the chlorosomes (Holt *et al.*, 1966; Broch-Due *et al.*, 1978). *Chloroflexus* resembles the purple bacteria in that oxygen inhibits Bchl synthesis (Pierson & Castenholz, 1974).

In conclusion, it may be said that phototrophic bacteria adapt to changes in light flux by changing the amount of the light-absorbing pigment complex in a largely inverse fashion. In *Rs. rubrum* this is strictly coordinated with changes in the amount of photochemical RC's. In other phototrophic bacteria, additional changes occur in the quantity of accessory LH complexes per reaction centre, leading to changes in the *in vivo* absorption properties (Fig. 2.2). In *Rp. sphaeroides*, and probably most other phototrophic bacteria, regulation of the accessory LH unit may be of major importance. This may be explained by the fact that water absorbs light in the near infrared (Pfennig, 1967) and that growth at deeper levels of shallow water bodies means growth at lower light fluxes, with less of this long-wavelength light. To a certain extent, this effect may be counteracted in most phototrophic bacteria by the observed production of accessory LH complexes absorbing at shorter wavelengths. At greater depths, however, light absorption is largely restricted to the carotenoids.

2.4.3 Formation of photosynthetically active membranes.

The early observations, from electron microscopical studies, that intracytoplasmic membranes are continuous with the cytoplasmic membrane (see Section 2.2) led to the hypothesis that the former arise by invagination of the latter (literature reviewed by Lascelles, 1968). Although this hypothesis was supported by biochemical investigations with *Rs. rubrum* (reviewed by Oelze & Drews, 1972), it was also questioned by Kaplan and Niederman and their co-workers (Kosakowski & Kaplan, 1974; Parks & Niederman, 1978; Fraley *et al.*, 1979a) who studied intracytoplasmic membrane formation in *Rp. sphaeroides*. These authors suggested that intracytoplasmic membrane vesicles may be independent of the cytoplasmic membrane in origin, structure and function, but that occasional secondary attachments might occur between the two types of membrane. More recent results, however, indicate that *Rs. rubrum* and *Rp. sphaeroides* do not necessarily follow the same pattern of intracytoplasmic membrane development. It is more likely that there are species specific differences. Additional investigations employing *Rs. tenue* have been of considerable help in the general understanding of the formation of photosynthetically active membranes.

At present, the following hypothesis appears to be well-supported by experimental data (Fig. 2.3): *Rs. tenue*, which does not form intracytoplasmic membranes, incorporates the photosynthetic apparatus homogeneously into the cytoplasmic membrane. *Rs. rubrum*, on the other hand, incorporates low but significant numbers of photosynthetic units into the cytoplasmic membrane which invaginates and thus forms intracytoplasmic membranes. Subsequently, the latter become the preferential sites for the further insertion of the photosynthetic apparatus. Finally, *Rp. sphaeroides* forms intracytoplasmic membranes by invagination and further differentiation at confined sites in the cytoplasmic membrane which differ from the unpigmented areas by the presence of Bchl complexes. Incorporation in particular of accessory LH Bchl (B800–850)-complexes then leads to an extension of the intracytoplasmic membrane system. These general conclusions can be derived from the following experimental results:

1. During the adaptive formation of the photosynthetic apparatus, *Rs. tenue* increased the area of peripheral membrane by altering its shape, i.e., increasing the surface area per cell (protein) (Wakim *et al.*, 1978). Freeze-fracture electron microscopy was employed to follow the production of the photosynthetic membrane, on the basis of the size and distribution of intra-membrane particles (Golecki & Oelze, 1980). The homogeneous

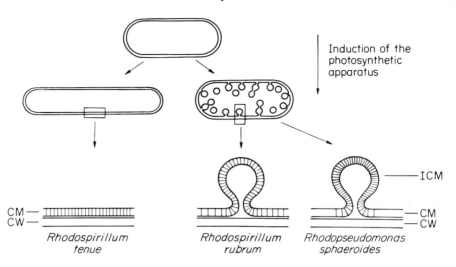

Fig. 2.3 Model of different modes of membrane differentiation in *Rhodospirillum tenue*, *Rhodospirillum rubrum*, and *Rhodopseudomonas sphaeroides*. CW = cell wall; CM = cytoplasmic membrane; ICM = intracytoplasmic membrane. Hatching represents the presence of bacteriochlorophyll-protein-complexes. For further details see text.

appearance over the entire membrane face of particles of characteristic size largely excludes at least in *Rs. tenue* the alternative possibility that photosynthetic membrane areas are formed and maintained independently of the existing membrane.

2. Cytoplasmic membrane preparations isolated from phototrophically grown *Rs. rubrum* always contain low but measurable amounts of Bchl. So far, there are no data to indicate that Bchl is incorporated into the cytoplasmic membrane at specific sites. During the adaptive formation of the photosynthetic apparatus, Bchl was detectable first in cytoplasmic membrane preparations and then predominantly in isolated chromatophore preparations derived from the newly-arising intracytoplasmic membranes. Pulse-chase experiments demonstrated that labelled phospholipids as well as labelled polypeptides could be traced from the cytoplasmic into the intracytoplasmic membrane. With respect to the development of intracytoplasmic membranes the following results were obtained. Initially, at low cellular Bchl levels, the specific Bchl content of chromatophore preparations increased. Subsequently, at cellular Bchl levels exceeding 10 nmol per mg protein, the specific content of Bchl (complexes) of chromatophore samples stayed constant. This means that in cells with more than 10 nmol of Bchl per mg cell-protein, alterations in the cellular content of Bchl reflect changes in the amount of intracytoplasmic membranes of fixed composition (reviewed by Oelze, 1981).

3. Two vesicle preparations of different composition could be obtained from the cytoplasmic membrane of *Rp. sphaeroides*, one of which contained Bchl and carotenoids while the other was unpigmented (Parks & Niederman, 1978; Niederman *et al.*, 1979). Radioactivity from the pigmented but not the unpigmented membrane fraction could be chased into chromatophores. Concomitantly the ratio of the accessory LH complex (B800–850) to the RC-LH unit (P_{865}–B875) increased (Niederman *et al.*, 1979). This suggests that intracytoplasmic membranes eventually become structures for housing predominantly the increasing amounts of LH (B800–850) complexes.

The formation of new membrane invaginations as well as the elongation of intracytoplasmic membranes would presumably be required in steady-state cultures to keep the cellular membrane content at a constant level. Elongation was suggested to proceed *via* the continuous incorporation of constituents like Bchl and protein over the entire surface of the intracytoplasmic membrane (Kosakowski & Kaplan, 1974; Broglie *et al.*, 1980).

2.4.4 Development of the photosynthetic apparatus in synchronously dividing populations of *Rhodopseudomonas sphaeroides*

Synchronized cell populations are required to study the dependency of various physiological parameters on the cell divisional cycle. Experiments on the formation of the bacterial photosynthetic apparatus by synchronously dividing phototrophic cultures of *Rp. sphaeroides* have been performed by S. Kaplan and his group.

In such cultures the proteins and pigments of the photosynthetic apparatus were inserted into cytoplasmic membranes at a constant rate throughout the cell cycle. Phospholipid incorporation, however, occurred mainly in a short period just before cell division (Fraley *et al.*, 1978b; Lueking *et al.*, 1978; Wraight *et al.*, 1978). [Similar observations have been made with non-phototrophic bacteria, e.g. *Escherichia coli*; Carty & Ingram, 1981]. Consequently, the protein to phospholipid ratio of the intracytoplasmic membranes increased to a maximum value and then fell just before cell division. The resulting cyclical changes amounted to 35–49 % of the protein to phospholipid ratio. Similar changes could be registered after determination of specific densities as well as of the fluidity of the lipid of isolated chromatophores (Fraley *et al.*, 1979a; b). It was proposed that these changes might serve as a means of regulating activities of membrane-bound enzymes. But because of the continuous insertion of various constituents like RC's, LH Bchl, as well as *b*- and *c*-type cytochromes (Wraight *et al.*, 1978) it seems unlikely that this type of regulation would include the entire formation of functional systems of the photosynthetic apparatus.

2.4.5 Development of light-dependent activities

The term photosynthesis describes the formation of stable products by activity of a reaction sequence started by light. This reaction sequence may be subdivided into the following functional units:

1. the LH unit, comprising the photopigments Bchl and carotenoids, which transform light energy into electronic energy;
2. the photochemical RC which transforms electronic energy into energy of charge separation by reducing the primary electron acceptor and oxidizing the primary electron donor;
3. secondary electron transport and proton translocation reactions which transform the energy of charge separation into energy of the electrochemical proton gradient; and
4. various reactions which utilize the energy of the electrochemical proton gradient for the production of energy-rich phosphate bonds, the reduction of ferredoxin and pyridine nucleotides or for transport processes.

In the following paragraphs, the development of the active photosynthetic apparatus will be described with reference to these functional units.

The development of functional LH units was studied with re-pigmenting cells of *Rp. sphaeroides* (Pradel et al., 1978; Hunter et al., 1979b). It was shown that, initially, the RC (P_{865}) LH (B875) Bchl units were incorporated separately at specific sites into the cytoplasmic membrane. Interestingly, the sites of early P_{865}–B875 incorporation were identified as membrane areas which serve as precursors in intracytoplasmic membrane formation (see Section 2.4.3). The efficiency of energy transfer increased as the amount of P_{865}–B875 units increased. Energy transfer between the units, however, became possible only at a relatively late stage when an enhanced incorporation of accessory (B800–850) LH complexes was detectable.

Information on the development of the secondary electron transport system was obtained by measuring the relative rates of light-induced proton extrusion by whole cells of *Rs. rubrum* (Oelze & Post, 1980). At low cellular Bchl levels, light-induced proton movement on a Bchl basis decreased as the cellular Bchl levels increased. At Bchl levels above about 10 nmol per mg of cell protein the specific rates of proton extrusion stayed largely constant.

Alterations in the rate of light-induced proton movement reflecting the activity of the secondary photochemical electron flow may be explained as follows: if, as discussed by several authors (see Jones, 1976), a section of the respiratory chain, including cytochromes, becomes involved in the photochemical electron transport, the amount of respiratory chain enzymes per photochemical reaction centre might directly influence the rate of photochemical electron flow. In turn this may be expected to influence the rates of reactions that depend on the electrochemical proton gradient. As a matter of fact, Keister & Minton (1969) showed that the rate of photophosphorylation

Table 2.3 Activities of membranes from selected species of the phototrophic bacteria grown phototrophically under different conditions.

Species	Growth cond.	Bacteriochlorophyll (nmol)			ATP (nmol) per min				NADH (nmol) oxidized	
		per Protein (mg)		Reaction centre (nmol)	per total Bchl (nmol)	Reaction centre (nmol)	Membrane protein (mg)		per total Bchl (nmol)	Membrane protein (mg)
		Cell	membrane							
Rs. rubrum*	4 (Wm^{-2})	22.7	65	25	1.4	35	43.5		0.33	21
	400	6.8	24	25	2.6	65	62		0.96	23
Rp. capsulata†	7 (Wm^{-2})	23.4	61	90.5	0.22	19.7	13		0.2	13
	2000	2.6	12	75	3.27	242	39		6.8	82
C. vinosum‡	33°C	55	142	190	0.73	139	104		—	—
	39°C	38	121	110	0.56	62	68		—	—

* Irschik & Oelze (1976).
† Schumacher & Drews (1979).
‡ Mechler & Oelze (1978b).

increased on a Bchl basis as the respiratory activity increased, in membranes from *Rs. rubrum* adapting from phototrophic to chemotrophic (high aeration) conditions in the dark. Table 2.3 shows that increased rates of photophosphorylation and respiration can be observed with membranes from cells of *Rs. rubrum* and *Rp. capsulata*, even when grown anaerobically at increased light energy fluxes.

The inverse relationship between Bchl contents and rates of photophosphorylation, however, does not appear to be general among the phototrophic bacteria. Thus, *Chromatium vinosum* shows a decreased rate of photophosphorylation as cellular Bchl contents decrease (Mechler & Oelze, 1978b) (Table 2.3).

One might assume that a phototrophic cell with larger amounts of membrane than a chemotrophic cell would contain more ATPase than the latter. This is not necessarily the case. The Ca^{2+}- and Mg^{2+}-dependent ATPase of membranes from *Rs. rubrum*, adapting from phototrophic to chemotrophic (highly aerobic) conditions, revealed increased activity on a Bchl basis as the cells' content of membranes and Bchl decreased (Post & Oelze, 1980). Thus, when expressed on a cell protein basis, the activity was constant.

The development of light-dependent NAD^+-reduction activity with succinate has only been studied with *Rs. rubrum*, as yet. On a Bchl basis, the rates were ten-times lower than those of photophosphorylation. But, apart from this quantitative difference, both rates changed in parallel (Keister & Minton, 1969; Irschik & Oelze, 1976).

In conclusion, the results reported on the development of light-dependent activities in membranes of the facultatively phototrophic bacteria investigated suggest that two distinct processes are involved:

1. The development of photosynthetic units, enabling efficient light harvesting and transfer of energy to RC's as well as from unit to unit. This process results in the extension of the entire photosynthetic membrane system.
2. The development of an active photochemical system with coupled secondary electron transport and related components.

In membranes of low Bchl content the second process is most pronounced, possibly because there is more membrane space per RC for these secondary components, especially those involved in the rate limiting step. The sum of these developmental processes in a phototrophically growing cell leads to a photosynthetic apparatus of such composition that it allows a largely stable rate of energy regeneration by the organisms irrespective of changes in the prevailing environmental conditions.

2.5 SUMMARY

The bacterial photosynthetic apparatus is localized in membranes which in only a few members are identical to the peripheral cytoplasmic membrane. In most of the species, the photosynthetically active membranes are intracytoplasmic, and exhibit tubular, vesicular or lamellar forms. Upon cell homogenization, intracytoplasmic membranes, particularly of the vesicular type, break down into closed vesicles known as chromatophores. Most of the studies concerned with the structure, function and development of the bacterial photosynthetic apparatus have been performed with chromatophore preparations.

The most prominent constituents of chromatophore membranes are carotenoids and bacteriochlorophylls, the latter in association with characteristic protein moieties. Lipid patterns are relatively simple, comprising phospholipids with fatty acids of chain lengths ranging predominantly from 14 to 18 carbon atoms. Remarkably, some members of the Rhodospirillineae exhibit a rather high proportion of mono-unsaturated fatty acids, while members of the Chlorobiineae produce glycolipids in addition to phospholipids. Most of the Bchl is contained in LH complexes and only minor amounts are associated with photochemical RC complexes which span the chromatophore membrane. One portion of the LH complexes of nearly all species is present at a constant ratio to the RC's. The other portion, which is not found in *Rs. rubrum*, varies independently of the reaction centre. As the two LH complexes exhibit different absorption properties, variations in their proportions in the near infrared are expressed by changes in the absorption properties of chromatophores and whole cells. These changes are a result of the regulation of Bchl synthesis which, within a relatively wide range, particularly affects the formation of the variable LH complex.

Currently, Bchl synthesis is assumed to be regulated by redox reactions, depending primarily on the energy-regenerating electron transport system, rather than by the cellular level or the rate of regeneration of energy-rich components. Bchl formation is the essential activity in the formation of photosynthetically active membranes which, in all organisms studied to date, is initiated by the differentiation, i.e., incorporation of Bchl complexes into the existing cytoplasmic membrane. However, the proportion of the cytoplasmic membrane which becomes involved in this differentiation ranges from the entire membrane, in *Rs. tenue*, to confined membrane regions, in *Rp. sphaeroides*. The function of the photosynthetic apparatus in facilitating energy-requiring reactions is reflected in facultative phototrophs by the ratio of respiratory activities per (RC) Bchl.

The rate of cellular energy regeneration is presumably kept at the required level by interplay of this latter effect with other factors. These include, variations in cellular membrane content and in Bchl content in response to variations in light energy supply.

Acknowledgements

The original investigations by the author have been financially supported by the Deutsche Forschungsgemeinschaft (Oe 53/2–5 and SFB 46).

REFERENCES

AAGAARD J. & SISTROM W. R. (1972) Control of synthesis of reaction center bacteriochlorophyll in photosynthetic bacteria. *Photochem. Photobiol.*, **15**, 209–25.

AMESZ J. (1978) Fluorescence and energy transfer. In *The Photosynthetic Bacteria* (Ed. by R. K. Clayton & W. R. Sistrom), pp. 333–40. Plenum Press, New York and London.

BACHOFEN R. (1979) Labeling of membranes and reaction centers from the photosynthetic bacterium *Rhodospirillum rubrum* with fluorescamine. *FEBS Lett.*, **107**, 409–12.

BIRRELL G. B., Sistrom W. R. & GRIFFITH O. H. (1978) Lipidprotein associations in chromatophores from the photosynthetic bacterium *Rhodopseudomonas sphaeroides*. *Biochemistry*, **17**, 3768–73.

BROCH-DUE M., ORMEROD J. G. & FJERDINGEN B. S. (1978) Effect of light intensity on vesicle formation in *Chlorobium*. *Arch. Microbiol.*, **116**, 269–74.

BROGLIE R. M., HUNTER C. N., DELEPELAIRE P., NIEDERMAN R. A., CHUA N.-H. & CLAYTON R. K. (1980) Isolation and characterization of the pigment-protein complexes of *Rhodopseudomonas sphaeroides* by lithium dodecyl sulfate/polyacrylamide gel electrophoresis. *Proc. natn. Acad. Sci. USA*, **77**, 87–91.

CARTY C. E. & INGRAM L. O. (1981) Lipid synthesis during the *Escherichia coli* cell cycle. *J. Bact.*, **145**, 472–8.

CLAYTON R. K. (1966) Spectroscopic analyses of bacteriochlorophylls *in vitro* and *in vivo*. *Photochem. Photobiol.*, **5**, 669–77.

CLAYTON R. K. & CLAYTON B. J. (1978) Molar extinction coefficients and other properties of an improved reaction center preparation from *Rhodopseudomonas viridis*. *Biochim. Biophys. Acta*, **501**, 478–87.

COGDELL R. J., PARSON W. W. & KERR M. A. (1976) The type, amount and energy transfer properties of the carotenoid in reaction centers from *Rhodopseudomonas sphaeroides*. *Biochim. Biophys. Acta*, **430**, 83–93.

COGDELL R. J., HIPKINS M. F. & SCHMIDT K. (1979) The effect of the light intensity on the light-harvesting pigment-protein composition of *Rhodopseudomonas acidophila*. Abstracts of the III. Int. Symp. on Photosynthetic Prokaryotes, Oxford, D 29.

COHEN-BAZIRE G. (1963) Some observations on the organisation of the photosynthetic apparatus in purple and green bacteria. In *Bacterial Photosynthesis* (Ed. by H. Gest, A. San Pietro & L. P. Vernon), pp. 89–114. Antioch Press, Yellow Springs, Ohio.

COHEN-BAZIRE G., SISTROM W. R. & STANIER R. Y. (1957) Kinetic studies of pigment synthesis by non-sulfur purple bacteria. *J. Cell. Comp. Physiol.*, **49**, 25–68.

DE BOER W. E. (1969) On the ultrastructure of *Rhodopseudomonas gelatinosa* and *Rhodospirillum tenue*. *Antonie van Leeuwenhoek, J. Microbiol. Serol.*, **35**, 241–2.

DREWS G. & OELZE J. (1980) Organization and differentiation of membranes of phototrophic bacteria. *Adv. Microb. Physiol.*, **22**, 1–91.

EIMHJELLEN K. E., STEENSLAND H. & TRÆTTEBERG J. (1967) A *Thiococcus* sp. nov. gen., its pigments and internal membrane system. *Arch. Mikrobiol.*, **59**, 82–92.

FEHER G. & OKAMURA M. Y. (1978) Chemical composition and properties of reaction centers. In *The Photosynthetic Bacteria* (Ed. by R. K. Clayton & W. R. Sistrom), pp. 349–86. Plenum Press, New York and London.

FENNA R. E. & MATTHEWS R. S. (1975) Chlorophyll arrangement in a bacteriochlorophyll protein from *Chlorobium limicola*. *Nature (Lond.)* **258**, 573–77.

FIRSOW N. N. & DREWS G. (1977) Differentiation of the intracytoplasmic membrane of *Rhodopseudomonas palustris* induced by variations of oxygen partial pressure or light intensity. *Arch. Microbiol.*, **115**, 299–306.

FOWLER C. F., NUGENT N. A. & FULLER R. C. (1971) The isolation and characterization of a photochemically active complex from *Chloropseudomonas ethylica*. *Proc. natn. Acad. Sci. USA*, **68**, 2278–82.

FRALEY R. T., JAMESON D. M. & KAPLAN S. (1978a) The use of the fluorescent probe α-parinaric acid to determine the physical state of the intracytoplasmic membranes of the photosynthetic bacterium *Rhodopseudomonas sphaeroides*. *Biochim. Biophys. Acta*, **511**, 52–69.

FRALEY R. T., LUEKING D. R. & KAPLAN S. (1978b) Intracytoplasmic membrane synthesis in synchronous cell populations of *Rhodopseudomonas sphaeroides*. *J. Biol. Chem.*, **253**, 458–64.

FRALEY R. T., LUEKING D. R. & KAPLAN S. (1979a) The relationship of intracytoplasmic membrane assembly to the cell division cycle in *Rhodopseudomonas sphaeroides*. *J. Biol. Chem.*, **254**, 1980–6.

FRALEY R. T., YEN G. S. L., LUEKING D. R. & KAPLAN S. (1979b) The physical state of the intracytoplasmic membrane of *Rhodopseudomonas sphaeroides* and its relationship to the cell division cycle. *J. Biol. Chem.*, **254**, 1987–91.

GARCIA A. F., DREWS G. & KAMEN M. D. (1975) Electron transport in an *in vivo*-reconstituted bacterial photophosphorylating system. *Biochim. Biophys. Acta*, **387**, 129–34.

GOLECKI J. R. & OELZE J. (1975) Quantitative determination of cytoplasmic membrane invaginations in phototrophically growing *Rhodospirillum rubrum*. *J. gen. Microbiol.*, **88**, 253–8.

GOLECKI J. R., BÜHLER R. & DREWS G. (1979) The size and number of intramembrane particles in cells of the photosynthetic bacterium *Rhodopseudomonas capsulata* studied by freeze-fracture electron microscopy. *Cytobiol.*, **18**, 381–9.

GOLECKI J. R. & OELZE J. (1980) Differences in the architecture of cytoplasmic and intracytoplasmic membranes of three chemotrophically and phototrophically grown species of the Rhodospirillaceae. *J. Bact.*, **144**, 781–8.

HOLT S. C. & MARR A. G. (1965a) Isolation and purification of the intracytoplasmic membranes of *Rhodospirillum rubrum*. *J. Bact.*, **89**, 1413–20.

HOLT S. C. & MARR A. G. (1965b) Effect of light intensity on the formation of intracytoplasmic membranes in *Rhodospirillum rubrum*. *J. Bact.*, **89**, 1421–9.

HOLT S. C., CONTI S. F. & FULLER R. C. (1966) Effects of light intensity on the formation of the photochemical apparatus in the green bacterium *Chloropseudomonas ethylicum*. *J. Bact.*, **91**, 349–55.

HUNTER C. N. & JONES O. T. G. (1979a) The incorporation of reaction centers into membranes from a bacteriochlorophyll-less mutant of *Rhodopseudomonas sphaeroides*. *Biochim. Biophys. Acta*, **545**, 325–38.

HUNTER C. N. & JONES O. T. G. (1979b) The kinetics of flash-induced electron flow in bacteriochlorophyll-less membranes of *Rhodopseudomonas sphaeroides* reconstituted with reaction centers. *Biochim. Biophys. Acta*, **545**, 339–51.

HUNTER C. N., VAN GRONDELLE R., HOLMES N. G., JONES O. T. G. & NIEDERMAN R. A. (1979b) Fluorescence yield properties of a fraction enriched in newly synthesized bacteriochlorophyll *a*-protein complexes from *Rhodopseudomonas sphaeroides*. *Photochem. Photobiol.*, **30**, 313–16.

IRSCHIK H. & OELZE J. (1976) The effects of transfer from low to high light intensity of electron transport in *Rhodospirillum rubrum* membranes. *Arch. Microbiol.*, **109**, 307–13.

JONES O. T. G. (1976) Electron transport and ATP Synthesis in the photosynthetic bacteria. *Symp. Soc. gen. Microbiol.*, **27**, 151–83.

KAISER I. & OELZE J. (1980) Growth and adaptation to phototrophic conditions of *Rhodospirillum rubrum* and *Rhodopseudomonas sphaeroides* at different temperatures. *Arch. Microbiol.*, **126**, 187–94.

KAPLAN S. (1978) Control and kinetics of photosynthetic membrane development. In *The Photosynthetic Bacteria* (Ed. by R. K. Clayton & W. R. Sistrom), pp. 809–39. Plenum Press, New York and London.

KEISTER D. L. & MINTON N. J. (1969) Interaction of photochemical and respiratory system of *Rhodospirillum rubrum*. *Progr. Photosynthesis Res.*, **III**, 1299–305.

KENYON C. N. (1978) Complex lipids and fatty acids of photosynthetic bacteria. In *The Photosynthetic Bacteria* (Ed. by R. K. Clayton & W. R. Sistrom), pp. 281–313. Plenum Press, New York and London.

KOSAKOWSKI M. H. & KAPLAN S. (1974) Topology and growth of the intracytoplasmic membrane system of *Rhodopseudomonas sphaeroides*: Protein, chlorophyll and phospholipid insertion into steady state anaerobic cells. *J. Bact.*, **118**, 1144–57.

LAMPE H. H., OELZE J. & DREWS G. (1972) Die Fraktionierung des Membransystems von *Rhodopseudomonas sphaeroides* und seine Morphogenese. *Arch. Mikrobiol.*, **83**, 78–94.

LASCELLES J. (1968) The bacterial photosynthetic apparatus. *Adv. Microb. Physiol.*, **2**, 1–42.

LASCELLES J. (1978) Regulation of pyrrole synthesis. In *The Photosynthetic Bacteria* (Ed. by R. K. Clayton & W. R. Sistrom), pp. 795–808. Plenum Press, New York and London.

LIEN S., GEST H. & SAN PIETRO A. (1973) Regulation of bacteriochlorophyll synthesis in photosynthetic bacteria. *Bioenergetics*, **4**, 423–34.

LÖW H. & AFZELIUS B. A. (1964) Subunits of the chromatophore membrane in *Rhodospirillum rubrum*. *Exp. Cell Res.*, **35**, 431–4.

LUEKING D. R., FRALEY R. T. & KAPLAN S. (1978) Intracytoplasmic membrane synthesis in synchronous cell populations of *Rhodopseudomonas sphaeroides J. Biol. Chem.*, **253**, 451–7.

MECHLER B. & OELZE J. (1978a) Differentiation of the photosynthetic apparatus of *Chromatium vinosum*, strain D. I. The influence of growth conditions. *Arch. Microbiol.*, **118**, 91–7.

MECHLER B. & OELZE J. (1978b) Differentiation of the photosynthetic apparatus of *Chromatium vinosum*, strain D. II. Structural and functional differences. *Arch. Microbiol.*, **118**, 99–108.

NIEDERMAN R. A., MALLON D. E. & PARKS L. C. (1979) Membranes of *Rhodopseudomonas sphaeroides*. IV. Isolation of a fraction enriched in newly synthesized bacteriochlorophyll a-protein complexes. *Biochim. Biophys. Acta*, **555**, 210–20.

NIETH K. F. & DREWS G. (1975) Formation of reaction center and light-harvesting bacteriochlorophyll-protein complexes in *Rhodopseudomonas capsulata*. *Arch. Microbiol.*, **104**, 77–82.

OELZE J. & DREWS G. (1972) Membranes of photosynthetic bacteria. *Biochim. Biophys. Acta*, **265**, 209–39.

OELZE J. & PAHLKE W. (1976) The early formation of the photosynthetic apparatus in *Rhodospirillum rubrum*. *Arch. Microbiol.*, **108**, 281–5.

OELZE J. & POST E. (1980) The dependency of proton extrusion in the light on the developmental stage of the photosynthetic apparatus in *Rhodospirillum rubrum*. *Biochim. Biophys. Acta*, **591**, 76–81.

OELZE J. (1981) Composition and development of the bacterial photosynthetic apparatus. *Subcell. Biochem.*, **8**, 1–73.

OLSON J. M., PRINCE R. C. & BRUNE D. C. (1977) Reaction-center complexes from green bacteria, Brookhaven Symposium on Biology, **28**, 238–45.

OYAMA H. & TUBOI S. (1979) Occurrence of a novel high molecular weight activator of δ-aminolevulinate synthetase in *Rhodopseudomonas sphaeroides*. *J. Biochem. (Tokyo)*, **86**, 483–9.

PARKS L. C. & NIEDERMAN R. A. (1978) Membranes of *Rhodopseudomonas sphaeroides*. V. Identification of bacteriochlorophyll a-depleted cytoplasmic membrane in phototrophically grown cells. *Biochim. Biophys. Acta*, **511**, 70–82.

PARSON W. W. (1978) Quinones as secondary electron acceptors. In *The Photosynthetic Bacteria* (Ed. by R. K. Clayton & W. R. Sistrom), pp. 455–69. Plenum Press, New York and London.

PFENNIG N. (1967) Photosynthetic bacteria. *Ann. Rev. Microbiol.*, **21**, 285–324.
PIERSON B. K. & CASTENHOLZ R. W. (1974) Studies of pigment and growth in *Chloroflexus aurantiacus*, a phototrophic filamentous bacterium. *Arch. Microbiol.*, **100**, 283–305.
PIERSON B. K. & CASTENHOLZ R. W. (1978) Photosynthetic apparatus and cell membranes of the green bacteria. In *The Photosynthetic Bacteria* (Ed. by R. K. Clayton & W. R. Sistrom), pp. 179–97. Plenum Press, New York and London.
POST E. & OELZE J. (1980) ATPase activities of chemotrophically and phototrophically grown *Rhodospirillum rubrum. FEMS Microbiol. Lett.*, **7**, 217–19.
PRADEL J., LAVERGNE J. & MOYA J. (1978) Formation and development of photosynthetic units in repigmenting *Rhodopseudomonas sphaeroides* wild type and 'phofil' mutant strain. *Biochim. Biophys. Acta*, **502**, 169–82.
PRINCE R. C., BACCARINI-MELANDRI A., HAUSKA G. A., MELANDRI B. A. & CROFTS A. R. (1975) Asymmetry of an energy transducing membrane. The location of cytochrome c_2 in *Rhodopseudomonas sphaeroides* and *Rhodoseudomonas capsulata. Biochim. Biophys. Acta*, **387**, 212–27.
REED D. W. & RAVEED D. (1972) Some properties of the ATPase from chromatophores of *Rhodopseudomonas sphaeroides* and its structural relationship to the bacteriochlorophyll proteins. *Biochim. Biophys. Acta*, **283**, 79–91.
REMSEN C. C. (1978) Comparative subcellular architecture of photosynthetic bacteria. In *The Photosynthetic Bacteria* (Ed. by R. K. Clayton & W. R. Sistrom), pp. 31–60. Plenum Press, New York and London.
SALTON M. R. J. & OWEN P. (1976) Bacterial membrane structure. *Ann. Rev. Microbiol.*, **30**, 451–82.
SANDY J. D., DAVIES R. C. & NEUBERGER A. (1975) Control of 5-aminolevulinate synthetase activity in *Rhodopseudomonas sphaeroides*: a role for trisulphides. *Biochem. J.*, **150**, 245–57.
SAUER K. & AUSTIN L. A. (1978) Bacteriochlorophyll-protein complexes from the light-harvesting antenna of photosynthetic bacteria. *Biochemistry*, **17**, 2011–19.
SCHACHMAN H. V., PARDEE A. B. & STANIER R. Y. (1952) Studies on the macromolecular organization of microbial cells. *Arch. Biochem. Biophys.*, **38**, 245–60.
SCHMIDT K. (1978). Biosynthesis of carotenoids. In *The Photosynthetic Bacteria* (Ed. by R. K. Clayton & W. R. Sistrom), pp. 729–50. Plenum Press, New York and London.
SCHMIDT K. (1980) A comparative study on the composition of chlorosomes (chlorobium vesicles) and cytoplasmic membranes from *Chloroflexus aurantiacus* strain Ok-70-fl and *Chlorobium limicola* f. *thiosulfatophilum* strain 6230. *Arch. Microbiol.*, **124**, 21–31.
SCHUMACHER A. & DREWS G. (1978) The formation of bacteriochlorophyll-protein complexes of the photosynthetic apparatus of *Rhodopseudomonas capsulata* during early stages of development. *Biochim. Biophys. Acta*, **501**, 183–94.
SCHUMACHER A. & DREWS G. (1979) Effects of light intensity on membrane differentiation in *Rhodopseudomonas capsulata. Biochim. Biophys. Acta*, **547**, 417–28.
SISTROM W. R. (1978) Control of antenna pigment components. In *The Photosynthetic Bacteria* (Ed. by R. K. Clayton & W. R. Sistrom), pp. 841–8. Plenum Press, New York and London.
STAEHELIN A., GOLECKI J. R., FULLER R. C. & DREWS G. (1978) Visualization of the supramolecular architecture of chlorosomes in freeze-fractured cells of *Chloroflexus aurantiacus. Arch. Microbiol.*, **119**, 269–77.
STRALEY S. C., PARSON W. W., MAUZERALL D. C. & CLAYTON R. K. (1973) Pigment content and molar extinction coefficients of photochemical reaction centers from *Rhodopseudomonas sphaeroides. Biochim. Biophys. Acta*, **305**, 597–609.
TAKEMOTO J. (1974) Kinetics of photosynthetic membrane assembly in *Rhodopseudomonas sphaeroides. Arch. Biochem. Biophys.*, **163**, 515–20.
TAKEMOTO J. & HUANG KAO M. Y. C. (1977) Effects of incident light levels on photosynthetic membrane polypeptide composition and assembly in *Rhodopseudomonas sphaeroides. J. Bact.*, **129**, 1102–9.

THORNBER J. P. & BARBER J. (1979) Photosynthetic pigments and models for their organization *in vivo*. In *Photosynthesis in Relation to Model Systems* (Ed. by J. Barber), pp. 27–70. Elsevier Biomedical Press, Amsterdam.

THORNBER J. P., TROSPER T. L. & STROUSE C. E. (1978) Bacteriochlorophyll *in vivo*: Relationship of spectral forms to specific membrane components. In *The Photosynthetic Bacteria* (Ed. by R. K. Clayton & W. R. Sistrom), pp. 133–60. Plenum Press, New York and London.

TUTTLE A. L. & GEST H. (1959) Subcellular particulate systems and the photochemical apparatus of *Rhodospirillum rubrum*. *Proc. natn. Acad. Sci. USA*, **45,** 1262–9.

UFFEN R. L. (1978) Fermentative metabolism and growth of photosynthetic bacteria. In *The Photosynthetic Bacteria* (Ed. by R. K. Clayton & W. R. Sistrom), pp. 857–72. Plenum Press, New York and London.

WAKIM B., GOLECKI J. R. & OELZE J. (1978) The unusual mode of altering the cellular membrane content by *Rhodospirillum tenue*. *FEMS Microbiol. Lett.*, **4,** 199–201.

WRAIGHT C. A., LUEKING D. R., FRALEY R. T. & KAPLAN S. (1978) Synthesis of photopigments and electron transport components in synchronous phototrophic cultures of *Rhodopseudomonas sphaeroides*. *J. Biol. Chem.*, **253,** 465–71.

Chapter 3. The Early Photochemical Events in Bacterial Photosynthesis

M. C. W. EVANS AND P. HEATHCOTE

3.1 INTRODUCTION

Clearly the definition of early events in photosynthesis depends very much on the viewpoint of the observer. In this chapter the early events are defined functionally as those which occur between the absorption of the energy of light quanta and the establishment of a stable reduced electron acceptor and oxidized electron donor in the electron transport chain of the bacterium. This definition also fits conveniently with the physical organization of the photosynthetic reaction centre which can be isolated containing all the components required for these early events, but without any of the secondary electron transport components. The subsequent electron transport reactions leading to ATP synthesis and the return of the electron from the low potential acceptor to the oxidized electron donor are dealt with in Chapter 4.

The main part of this chapter deals with the purple bacteria, with a shorter section describing the green bacteria. The purple bacteria are better understood because of the ease with which they can be isolated and grown. This in turn has made possible large-scale biochemical studies and led to the development of genetic studies and isolation of useful mutants. It has also become apparent that the purple bacteria have the simpler and biochemically more accessible photochemical system. The green bacteria on the other hand are more difficult to grow and their adaptation to growth at low light intensities has led to a very complex physical organization of the photosynthetic apparatus, making biochemical and biophysical work more difficult.

3.2 THE PURPLE BACTERIA

3.2.1 What are the photosynthetic pigments in phototrophic bacteria?

The light absorbing pigments in phototrophic bacteria can be easily identified by measuring the absorption spectrum of the cells. Figs. 3.1a and 3.1b show the absorption spectra of a purple bacterium *Rhodopseudomonas sphaeroides* and a green bacterium, *Chlorobium limicola* f. *thiosulfatophilum*. These spectra show major absorbance peaks in the 400 nm region and between 700 and 900 nm which are due to the absorption of bacteriochlorophyll *a* in *Rps. sphaeroides* and mainly of bacteriochlorophyll *c* in *C. limicola* (see Chapter 1 for structure of the chlorophylls). Fig. 3.1 also shows the spectra of these chlorophylls extracted into methanol. There are clearly large differences between the *in vivo* and *in vitro* spectra, caused by the binding of the chlorophylls to specific proteins *in vivo* to form the light-harvesting and reaction centre complexes. The spectra of the intact cells also show considerable absorption in the 500 nm region, particularly in *Rp. sphaeroides*. This is due to carotenoid protein complexes. A number of mutant strains have been

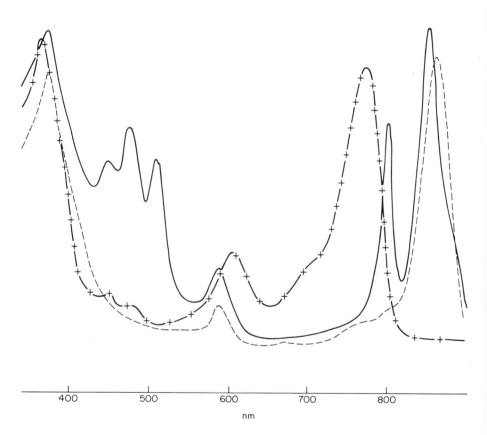

Fig. 3.1 The absorption spectra of green and purple phototrophic bacteria.
(a) ——— *Rp. sphaeroides* wild type.
– – – – *Rp. sphaeroides* Str. R.26.
—+—+ Bacteriochlorophyll *a* in methanol extract of *Rp. sphaeroides*.

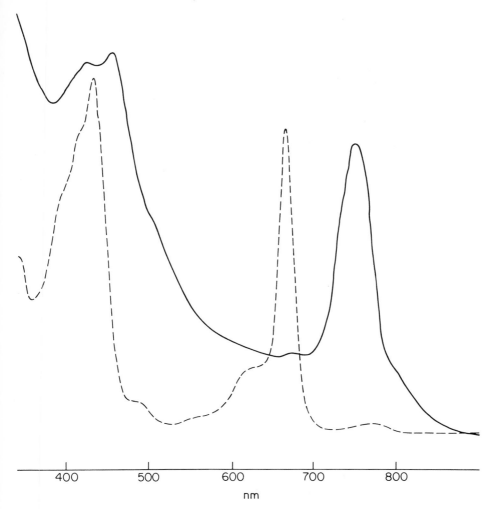

Fig. 3.1(b) ——— *Chlorobium limicola* f. *thiosulfatophilum.*
――――― Methanol extract of bacteriochlorophyll *c* and *a* from *C. limicola* f. *thiosulfatophilum.*

isolated which lack carotenoids. They are blue-green and have been very useful in making measurements of photosynthetic reactions by absorption spectroscopy, as the carotenoids normally mask many of the spectral changes of cytochromes and other components. Fig. 3.1 also shows the absorption spectrum of *Rp. sphaeroides* Str. R.26, one of the most commonly used carotenoid-less mutants. This strain also lacks the major bacteriochlorophyll protein absorbing at 800 nm.

The relationship between the absorbing pigments and photosynthetic activity can be found by making an action spectrum in which the effectiveness of light of different wavelengths in activating photosynthesis is measured. The number of quanta absorbed is determined and correlated with an activity, which may be overall growth or a specific photochemical event, to determine the quantum efficiency. If this is done with phototrophic bacteria it is found that light absorbed by the bacteriochlorophyll protein complexes is used with high efficiency and light absorbed by the carotenoids with an efficiency that varies between 30 and 90% in different species. The bacteriochlorophyll proteins are, then, the major photosynthetic pigments, with the carotenoids acting as accessory pigments to extend the usable wavelength range. The efficiency with which energy absorbed by carotenoids is used may depend on how closely the carotenoid is associated with the bacteriochlorophyll protein. Carotenoids also function under aerobic conditions to protect the photosynthetic apparatus from damage by singlet oxygen produced as a result of the interaction of an excited pigment molecule with oxygen.

3.2.2 How is the light energy trapped?

Photosynthesis is a highly organized photochemical oxidation-reduction process. The primary energy conserving reaction is the photo-oxidation of a chlorophyll molecule (Fig. 3.2). When a chlorophyll molecule absorbs a quantum of light it is transformed into a singlet excited state in which one electron is raised to a higher energy level:

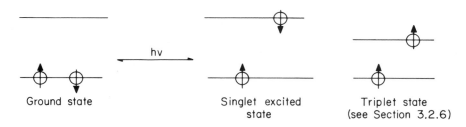

Ground state   Singlet excited   Triplet state
                         state   (see Section 3.2.6)

In the excited state the properties of the molecule are altered and it can act as a reducing agent at a much lower redox potential than the ground state molecule. If a suitable electron acceptor is available it becomes reduced by the excited chlorophyll. The redox potential difference between the reduced acceptor and the oxidized chlorophyll molecule represents the photochemically trapped energy. If a suitable mechanism for reducing the oxidized chlorophyll exists, a photochemical redox cycle can be established. The operation of such a cycle as the energy conserving mechanism in photosynthesis was first suggested by *in vitro* experiments in which solutions of

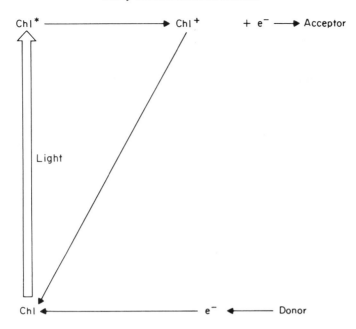

Fig. 3.2 Diagrammatic representation of the photo-oxidation of chlorophyll. Chl: Chlorophyll   Chl*: singlet excited state of Chlorophyll.

chlorophyll in an organic solvent catalysed photochemical redox reactions. In solution such reactions show poor quantum efficiencies because the excited state may lose energy rapidly by non-radiative transition or by fluorescence. The fluorescence lifetime of most of the bacteriochlorophyll in purple bacteria is about 10 ns, so very efficient oxidation systems have evolved in photosynthesis to allow high quantum efficiencies.

3.2.3 The photosynthetic unit

On the supposition that bacteriochlorophyll is the main light-absorbing pigment, and that it can undergo photochemical redox reactions, it is reasonable to assume that it is the photoactive pigment in photosynthesis. The phototrophic bacteria do indeed contain large amounts of chlorophyll, up to 20% of the cellular dry weight in some *Chlorobium* spp. Each of these chlorophyll molecules could in theory act as a photoredox centre. However, this would require a complete complement of electron donors and acceptors for each of them. This would be rather inefficient as the quantum fluxes in the environments in which green and purple bacteria grow are usually quite low, such that each chlorophyll molecule will receive a quantum only at long time intervals, perhaps once every 5–10 minutes on average. A more effective mechanism might be to have a large number of chlorophyll molecules collaborating to provide electrons for a smaller number of electron transport chains. That this is in fact the case was first shown with green algae. When algae are illuminated with short flashes of light of varying intensity the maximum rate of photosynthesis is observed not when each flash contains enough quanta to provide one quantum per chlorophyll, but when only one quantum per 300 chlorophylls is available in each flash. These experiments led to the concept of the photosynthetic unit (see Section 2.4).

A photosynthetic unit consists of a large number of light absorbing pigment molecules associated with a single photochemical reaction centre and a single electron transport chain. Within a photosynthetic unit a quantum of light may be absorbed by any pigment molecule, which is thereby converted into its excited state. However, it does not undergo photo-oxidation; instead, the trapped energy moves by non-radiative inductive resonance transfer to an adjacent pigment molecule. This process continues randomly until either the energy is lost as heat or fluorescence, or is trapped by the photochemical reaction centre. Only in the latter case is the energy photosynthetically useful.

The amount of energy required to excite a chlorophyll molecule *in vivo* is controlled by its protein environment. The minimum energy required is indicated by the longest wavelength absorption band of the chlorophyll protein complex. The photochemical reaction centre contains bacteriochlorophyll in an environment which requires a lower energy for excitation than in the case of the antenna pigment. The reaction centre chlorophyll can be detected by its longer wavelength absorption maximum. When energy is transferred to the reaction centre from the light-harvesting chlorophyll a small part is lost as heat. Energy cannot then be easily transferred back to the light-harvesting chlorophyll. The excited reaction centre can then be photo-oxidized or the energy lost by fluorescence. The maximum energy available for photosynthesis is determined by the absorption band of longest wavelength of the reaction centre chlorophyll. The extra energy available from quanta absorbed at higher energies (shorter wavelength) by other absorption bands of the chlorophyll is lost as heat by intramolecular decay to the lowest excited singlet state. The high quantum efficiencies that can be observed with biological systems (Wraight & Clayton, 1973) indicate that the structure of the photosynthetic unit and the ratio of light-harvesting pigment molecules to reaction centres has evolved to give a very high probability of energy transfer to the reaction centre.

In general the size of the photosynthetic unit can be varied in response to environmental changes to optimize light harvesting. In most purple bacteria there may be 50–500 light harvesting bacteriochlorophyll molecules per reaction centre. There is less variation in green sulphur bacteria which usually contain 100 bacteriochlorophyll *a* and 1–2000 bacteriochlorophyll *c*, *d* or *e* per reaction centre. In oxygenic organisms there are two different types of reaction centre with 300 chlorophylls per reaction centre, but one electron transport chain linking the two reaction centres.

The organization of the light-harvesting complex of purple bacteria is dealt with in Chapter 2. Models of the organization of the light-harvesting proteins in the membrane of purple bacteria suggest that B875 surrounds and is closely associated with the reaction centre while B800–850 forms a variable layer outside this. Such a model also implies that energy from the B800–850 is transferred to the reaction centre through the B875 (Cogdell & Thornber, 1979).

3.2.4 Isolation and characterization of the reaction centre

Biophysical studies have shown that the reaction centre functions as an energy trap and biochemical studies reveal that it exists as a defined physical entity in the membrane, organized to provide optimal conditions for photosynthetic energy conversion. The work on reaction centres originated mainly in the laboratory of Clayton and his co-workers who developed the use of detergents, and mutants lacking non-essential parts of the apparatus such as carotenoids, notably the R26 strain of *Rp. sphaeroides* (Reed & Clayton, 1968), to isolate the reaction centre as a defined component.

Reaction centres have been isolated with varying degrees of purity from a number of purple bacteria including *Rp. capsulata*, *Rhodospirillum rubrum*, *Chromatium vinosum* and the bacteriochlorophyll (Bchl) *b* containing organisms *Rp. viridis* and *Thiocapsa pfennigii*. However, much the most researched and best characterized reaction centres are those from *Rp. sphaeroides*, particularly those from the R26 mutant (see review by Feher & Okamura, 1978).

This reaction centre preparation has three polypeptides of 28 (H), 24 (M) and 21 (L) kilodaltons. It contains 4 bacteriochlorophylls, 2 bacteriopheophytins, 2 ubiquinone molecules and one iron atom. Some of these components, including the H subunit, the iron and one quinone, can be removed without loss of the essential primary photochemistry. However, the kinetics of the reactions are affected, indicating that the complete unit is required for optimal activity. Reaction centres from wild-type organisms contain in addition a carotenoid molecule, and reaction centres from some organisms, like *Chr. vinosum* and *Rp. viridis*, a *c* type cytochrome.

3.2.5 The mechanism of energy conservation in the reaction centre

Fig. 3.3 shows diagrammatically the reactions that occur in the reaction centre following excitation by a quantum of light. The reaction centre chlorophyll P_{870} absorbs the energy of the quantum and is raised to an excited state. In this

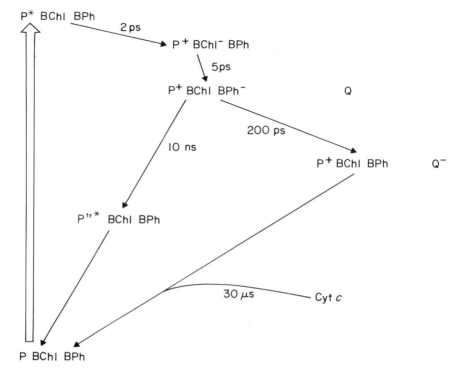

Fig. 3.3 Diagrammatic representation of the electron transfer reaction in the purple bacterial reaction centre.

P: the reaction centre bacteriochlorophyll P_{870}.
BChl: The B800 bacteriochlorophyll.
BPh: bacteriopheophytin.
Q: the primary quinone electron acceptor.

state it is a strong reducing agent and transfers an electron via one of the bacteriochlorophylls with an absorption at 800 nm to one of the bacteriopheophytin molecules and then on to a bound quinone. These reactions have to occur very quickly to compete with energy loss by the fluorescence from the pigments. During the 1970s techniques were developed using very short (picosecond, 10^{-12} s) pulses of light from lasers, and light detection systems capable of picosecond time resolution for the measurement of absorption changes in this time range. Application of these techniques to photosynthesis (see review by Blankenship & Parson, 1978) has shown that electron transfer from P_{870} to the bacteriopheophytin intermediate occurs in about 10 ps and from the intermediate to the quinone in about 150 ps. Recent work by Shuvalov & Parson (1981) indicates that one of the B800 bacteriochlorophyll molecules in the reaction centre is also involved in accepting an electron from the P_{870} in less than 3 ps. If the reaction centre is blocked by pre-reduction of the quinone the energy is lost from the bacteriopheophytin by recombination of the excited electron with the P_{870} through a triplet state of P_{870} (see Section 3.2.2) and delayed emission of the energy as a photon of light after a few nanoseconds. The energy (redox) level of the excited reaction centre chlorophyll, the reduced B800 and bacteriopheophytin are very similar, allowing rapid back reactions, while the quinone electron acceptor has a much lower energy (redox) level and transfer of the electron down to this acceptor stabilizes the charge separation into the time domain of normal electron transport reactions.

3.2.6 Identification and properties of the reaction centre components

The electron donor, reaction centre chlorophyll

Early experiments on the absorption spectrum of the reaction centre indicated that the photo-oxidizable component was probably a bacteriochlorophyll in an environment different from that of the light-harvesting chlorophyll. This was confirmed by the isolation of reaction centres from *Rp. sphaeroides* R26. The only pigments in these were bacteriochlorophyll and bacteriopheophytin. There is an absorption peak at about 870 nm (960 nm in Bchl *b* containing organisms) which is bleached on illumination in the presence of an electron acceptor or on chemical oxidation (Fig. 3.4). There are also changes in the 590 nm peak and in the 400 nm region of the bacteriochlorophyll absorbance. These changes clearly indicated that the photo-oxidizable component is a bacteriochlorophyll. In the time scale of steady state spectroscopic measurements no changes are observed in the B800 bacteriochlorophyll or bacteriopheophytin.

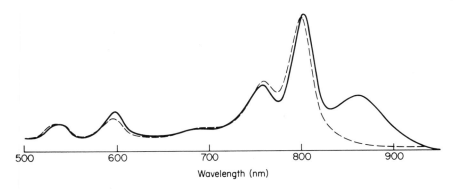

Fig. 3.4 The absorption spectrum of the isolated reaction centre of *Rp. sphaeroides* Str. R.26.
———— Reduced. — — — — Oxidized.

Table 3.1 Midpoint oxidation–reduction potentials of reaction centre components.

	Rp. sphaeroides	Rp. viridis	Chromatium vinosum	Chlorobium limicola
P-P$^+$	+450	+475	+490	+250
Bph-Bph$^-$	< −600	−620 (−400)		
Q_1-Q_1^- indirect	−20	−95	−100	−550*
direct	−45	−120	−90	
Q_1^--Q_1H_2	[−550]	−525		
Q_2-Q_2^-	+100	+130		
Q_2^--Q_2H_2	+20	+30		

The values given are based on measurements made by many authors. For comprehensive references see Dutton & Prince (1978) and Rutherford & Evans (1980).

P: the reaction centre bacteriochlorophyll
Bph: bacteriopheophytin intermediate
Q_1: primary quinone electron acceptor
Q_2: secondary quinone electron acceptor
(Values for Q_1 and Q_2 are at pH 7.0)

* The primary acceptor in *C. limicola* is probably an iron-sulphur protein, not a quinone.

The oxidation of the reaction centre chlorophyll can also be observed by electron paramagnetic resonance spectrometry (EPR). This technique detects molecules with unpaired electrons. It is particularly useful in studies of photosynthesis because it allows the detection of one-electron redox reactions such as the photo-oxidation of the reaction centre. One of the earliest observations of reaction centre photo-oxidation was in the experiments of Commoner and his colleagues (Commoner *et al.*, 1956) who not only observed a light-induced EPR signal but also showed that the photochemical reaction occurred in frozen samples at temperatures as low as 4 K. The ability of the photochemical system to operate at this low temperature provides strong evidence for the highly organized nature of the reaction centre and that electron transfer proceeds by an electron tunnelling mechanism (Hopfield, 1974). The light-induced EPR signal is different from the EPR signal of oxidized bacteriochlorophyll in solution, being narrower by a factor of about $\sqrt{2}$ (Fig. 3.5). This suggests that the single, unpaired electron is shared between two chlorophyll molecules, i.e. it is a dimer of bacteriochlorophyll molecules that forms the reaction centre. This proposal is confirmed by measurements using a second magnetic resonance technique, electron nuclear double resonance. This allows a more precise determination of the distribution of the electron and confirms that it is shared between two molecules. This type of dimeric bacteriochlorophyll molecule cation would be expected to have an absorption band in the near infrared around 1250 nm when it is reduced, and such a band with kinetic characteristics parelleling those of the 870 nm band has been observed (Dutton *et al.*, 1977). As mentioned above, when the quinone acceptor is reduced prior to illumination the energy trapped in the charge separation between the P_{870} and the pheophytin intermediate can be

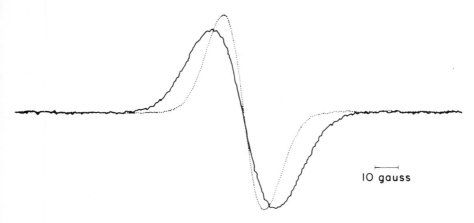

Fig. 3.5 The EPR spectra of oxidized bacteriochlorophyll *a*.
——— Oxidized monomeric bacteriochlorophyll *in vitro*.
······ Photo-induced signal in *Rhodospirillum rubrum* of the dimeric bacteriochlorophyll of the reaction centre.
(Reproduced from Norris & Katz (1978) with permission).

lost through a P_{870} triplet state. This triplet state can be detected optically or by EPR and has unusual properties. Its spin polarization characteristics, indicated by the symmetry of emission and absorption lines in the low temperature EPR spectrum, suggest that it is formed by a recombination of P_{870}^+ and the reduced pheophytin as outlined above, rather than by decay of the P_{870} excited singlet. Its zero-field splitting characteristics, determined from the EPR spectrum, also show that it could not arise from a monomeric bacteriochlorophyll, thus offering more support for the dimer model of the reaction centre (Norris & Katz, 1978). Studies of the optical properties of the reaction centre chlorophyll in oriented membrane preparations indicate that the porphyrin rings of the dimer are parallel to each other and to the plane of the membrane.

The redox potential of the P_{870} ground state has been determined by a number of experiments to be about $+450$ mV.

The intermediary electron carrier

The intermediary electron acceptor was first identified by investigation of time-resolved spectra in the picosecond time range following excitation of the reaction centre by a short light pulse. At times longer than about 200 ps the absorption changes observed in the 400–900 nm region of the spectrum are those of the oxidation of P_{870} or P_{960} (Fig. 3.6a). At shorter times (10–20 ps) the spectrum is quite different and is a mixture of the oxidized P_{870} spectrum and the reduced spectrum of bacteriopheophytin (Fig. 3.6a). In the 20–200 picosecond time range the bacteriopheophytin part of the spectrum decays as

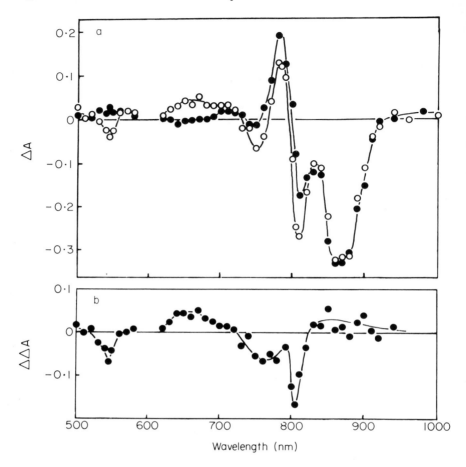

Fig. 3.6 (a) The difference spectra of the absorbance changes in reaction centres of *Rp. sphaeroides* following excitation with a 7 ps laser flash at 600 nm measured after 20 ps (open circles) and 3 ns (closed circles).

(b) The spectrum of the intermediary electron carrier (bacteriopheophytin) obtained as the difference between the two spectra shown in (a).

(Reproduced from Schenck *et al.*, 1981 with permission).

Fig. 3.7 EPR spectrum of the bacteriopheophytin intermediary electron carrier in *Rp. viridis*.

(a) Spectrum recorded at 77 K. (b) Spectrum recorded at 8 K.

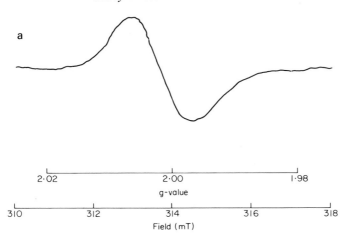

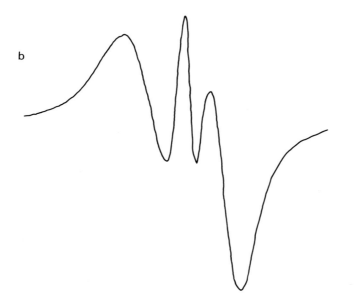

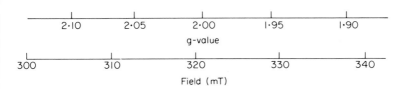

Fig. 3.7

the electron is transferred on to the quinone electron acceptor. The spectrum of the bacteriopheophytin intermediate can be obtained without interference from P_{870}^+ by trapping the intermediate in the reduced form by illumination of samples prepared at low redox potential. The same spectrum is obtained as the difference between the flash induced spectra, Fig. 3.6b. This spectrum matches that of the chemically reduced bacteriopheophytin (Fajer *et al.*, 1977). The reduced pheophytin can also be detected by EPR. Under most conditions the spectrum is a symmetrical 13 gauss-wide signal at $G = 2.003$ (Fig. 3.7a), essentially the same as that of the reduced monomeric bacteriopheophytin in solution. However, at very low temperatures this signal is split to give a 60 gauss-wide doublet (Fig. 3.7b). This splitting is thought to be due to magnetic interaction between the pheophytin anion, the semi-quinone anion of the quinone acceptor and an adjacent ferrous iron atom. It is an important indicator of the distances between the components.

Determination of the *in vivo* redox potential of the pheophytin intermediate has proved difficult, with reported values around -400 mV (Prince *et al.*, 1976) and about -600 mV (Klimov *et al.*, 1977, Rutherford *et al.*, 1979). The *in vitro* potential is -580 mV so the lower reported potential *in vivo* is perhaps correct.

The quinone electron acceptors

After the initial charge separation and the reduction of the bacteriopheophytin intermediary electron carrier, the charge separation is stabilized by transfer of the electron to the primary quinone electron acceptor. This electron transfer results in a considerable loss of energy, since the redox potential of the quinone is about -180 mV, compared to the pheophytin potential of about -600 mV. This energy may be available to the bacterium as a membrane potential (see Chapter 4) or it may be sacrificed to ensure that the reaction proceeds in the correct direction.

The quinone acceptor was first identified as an EPR detectable component undergoing photoreduction in parallel with the oxidation of P_{870} at low temperature (Leigh & Dutton, 1972). The EPR signal is a broad, rather unusual signal near $G = 1.82$ in most species (Fig. 3.8). It is not a characteristic semiquinone signal. However, if reaction centres are prepared under conditions such that they lose the iron atom normally associated with them, this signal is replaced by a characteristic semiquinone signal at $G = 2$. The $G = 1.82$ signal is thought to arise as a result of magnetic interaction between the semiquinone anion and the iron atom in the reaction centre.

When the electron transport chain is operating at room temperature the electron is rapidly transferred to a secondary quinone acceptor. During normal functioning the primary quinone operates between the fully oxidized and the semiquinone forms; it is never fully reduced and it is not protonated, so that its functional redox potential is that at the pK, about -180 mV. The

Fig. 3.8 EPR spectra of quinone-iron electron acceptors in *Rp. sphaeroides*.
(a) The primary quinone acceptor chemically reduced to the semiquinone.
(b) The primary quinone acceptor reduced by illumination at 7 K.
(c) The secondary quinone acceptor chemically reduced to the semiquinone.
Spectra were recorded at 7 K.

secondary quinone acceptor acts as a two-electron gate. When both quinones are oxidized, a single turnover of the reaction centre results in oxidation of P_{870} and reduction of the secondary quinone to the semiquinone. This semiquinone is very stable and does not function as an electron donor to the electron transport chain. It is only after a second turnover of the reaction centre when the secondary quinone is fully reduced and protonated that two electrons are donated rapidly to the electron transport chain.

The chromatophore membrane contains about 20–30 quinone molecules per electron transport chain. However, experiments in which most of the quinone is extracted by organic solvents suggest that only three quinone molecules are required for electron transport, two of these being the reaction centre quinones. These appear to be specifically bound to proteins in the reaction centre. In some organisms they are ubiquinone, which is also the bulk quinone, but in others such as *Chr. vinosum* and *Rp. viridis* they are menaquinone.

The EPR spectrum of the secondary quinone is similar to that of the primary quinone indicating that it also interacts with the iron atom (Fig. 3.8). When both quinones are in the semiquinone state no EPR spectrum is detectable. This is because of magnetic interaction between the two semi-quinones and the iron atom and it indicates that they are not more than 20 Å apart.

The secondary quinone is readily lost from reaction centres during preparation, whereas the primary quinone is much more tightly bound, but can be removed in order to confirm its role as electron acceptor. Depleted reaction centres can be reconstituted with quinone. When a modified quinone which works as a photo-affinity label is used, it can be shown that the quinone binding site is on the M peptide (Marinetti *et al.*, 1979).

The electron donor to P_{870}

In all purple bacteria the electron donor to P_{870} is a *c* type cytochrome. In some species this cytochrome is an integral part of the reaction centre and will donate electrons to P_{870} even at low temperature (10 K). In others it is a soluble cytochrome easily lost during purification.

3.2.7 Structure and function of the reaction centre

It is not yet possible to provide a complete model of the reaction centre. A number of indications of the basic structure are available. The three polypeptide subunits are present in a 1:1:1 ratio.

A transmembrane organization of the reaction centre photochemistry is indicated by a number of functional experiments. There is evidence that charge separation across the membrane is associated with the primary photochemistry. The cytochrome c which donates electrons to P_{870} is located on the periplasmic side of the membrane, and proton binding to the secondary quinone acceptor occurs on the cytoplasmic side of the membrane (Dutton & Prince, 1978).

Magnetic resonance experiments provide some information on the proximity of the reaction centre components. Strong magnetic interaction effects on the EPR spectra of both quinones and the bacteriopheophytin indicate that all are within about 10 Å of the iron atom. Recent measurements using the electron spin echo technique also show weak interaction of P_{870} with the iron atom suggesting that the distance between them is less than 20 Å (Bowman *et al.*, 1979). These measurements indicate that the photochemically active components of the reaction centre occupy a space not more than 30 Å wide

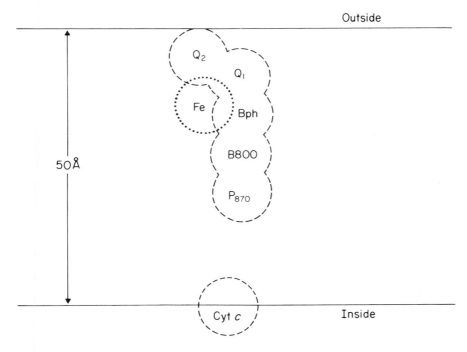

Fig. 3.9 A model of the purple bacterial reaction centre indicating the spacing of the active centres of the electron transfer components based on their magnetic interactions.
P_{870}: reaction centre bacteriochlorophyll.
Bph: bacteriopheophytin.
B800: bacteriochlorophyll.
Q_1 and Q_2: Primary and secondary quinone electron acceptors.
Cyt: cytochrome.
Fe: reaction centre associated ferrous iron atom.
The terms 'outside' and 'inside' refer to the chromatophore membrane and signify the opposite of the sidedness *in vivo*.

within the membrane which is 50–80 Å thick. Fig. 3.9 shows diagrammatically the required arrangement of the reaction centre.

3.3 THE GREEN PHOTOTROPHIC BACTERIA

These can be divided into two groups, the green sulphur bacteria, e.g. the genus *Chlorobium* and the green gliding filamentous bacteria, e.g. the genus *Chloroflexus*. The light-harvesting apparatus of the green bacteria has been characterized, and differs markedly from that of the purple bacteria. The structure of the light-harvesting apparatus is more complicated, and includes bacteriochlorophylls *c*, *d* or *e* as well as bacteriochlorophyll *a*. Relatively little is known about the reaction centres of green bacteria because they have not been separated from this complex system. However it has been shown that the green sulphur bacteria contain a photosynthetic reaction centre that resembles photosystem I of higher plants and cyanobacteria rather than that of the purple bacteria.

3.3.1 The light-harvesting apparatus

Green bacteria contain bacteriochlorophyll *a* and larger amounts of bacteriochlorophylls *c*, *d* or *e*. The most important difference between bacteriochlorophyll *a* and bacteriochlorophylls *c*, *d* or *e* is that ring II in bacteriochlorophyll *a* is more reduced (see Chapter 1). As a result the main absorption bands of bacteriochlorophyll *a in vivo* lie below 400 nm or above 800 nm, whereas those of bacteriochlorophyll *c*, *d* or *e* lie above 400 nm and below 800 nm. The absorption spectrum of whole cells of the green sulphur bacterium *Chlorobium limicola* f. *thiosulfatophilum* is shown in Fig. 3.1b. Bacteriochlorophyll *c* is responsible for the absorption bands at 460 nm and 750 nm, whilst bacteriochlorophyll *a* contributes the minor absorption band at 810 nm. The photochemically active reaction centre bacteriochlorophyll, which is thought to be bacteriochlorophyll *a*, bleaches at 840 nm upon oxidation and is called P_{840}.

The green sulphur bacteria are strict anaerobes and obligate photoautotrophs that grow in relatively dim light. They have much larger photosynthetic units than those found in purple bacteria. Depending on the light intensity during growth up to 1000–2000 bacteriochlorophyll *c*, *d* or *e* molecules and approximately 100 bacteriochlorophyll *a* molecules are associated with each reaction centre. The green, gliding filamentous bacteria are facultative photoautotrophs that grow well at relatively high light intensities. They appear to have smaller photosynthetic units than the green sulphur bacteria, but the specific cellular Bchl *c* content can vary at least 30-fold depending upon the light intensity (Pierson & Castenholtz, 1974).

Electron microscopy of thin sections of green bacteria show that they do not possess the unit membrane vesicles that contain the photosynthetic apparatus in purple bacteria. Instead the green bacteria contain oblong bodies (30–70 nm × 100–260 nm) adpressed to the inside of the cytoplasmic membrane. These structures were originally called 'chlorobium vesicles'. However, since they are not bounded by a unit membrane, and are not restricted to the genus *Chlorobium*, they have been renamed 'chlorosomes'. Fractionation studies have indicated that essentially all of the bacteriochlorophyll c, d or e and most of the carotenoids are localized in the 'chlorosomes', while bacteriochlorophyll a, some carotenoid, and bacteriopheophytin c are located in the cytoplasmic membrane. The cytochromes of the electron transport chain are associated with the cytoplasmic membrane, indicating that this membrane contains the photosynthetic reaction centres.

Two types of light-harvesting bacteriochlorophyll a proteins are associated with the reaction centre and cytochromes in the cytoplasmic membrane. Approximately half of the bacteriochlorophyll a is contained in a water-soluble pigment-protein, which was shown to be only loosely associated with the reaction centre. This protein complex when isolated has major absorption bands at 809, 604 and 373 nm. It is the only photosynthetic pigment-protein to have been crystallized *in vitro*, and has been studied by X-ray diffraction (Matthews *et al.*, 1979). Both in solution and as crystals the water-soluble bacteriochlorophyll a protein isolated from the green sulphur bacterium *Prosthecochloris aestuarii* exists as a trimer of three pigment-protein subunits. Each subunit contains seven bacteriochlorophyll a molecules wrapped in a polypeptide chain, with an average centre-to-centre distance of 1–2 nm. As the water-soluble bacteriochlorophyll a protein isolated from *C. limicola* has the same polypeptide subunit size, and similar absorption and circular dichroism (CD) spectra as that isolated from *P. aestuarii*, the two pigment-proteins probably have the same structure. Water-insoluble bacteriochlorophyll a proteins have been found to be more closely associated with the reaction centre. An extremely elegant freeze-fracture electron microscopy study by Staehelin *et al.* (1980) of *C. limicola* has described the structural organization of the pigments (Fig. 3.10). The chlorosomes are bounded by a 2–4 nm wide envelope layer, possibly consisting of a galactolipid monolayer as chlorosomes are rich in galactolipids. Each chlorosome contains 10–30 rod elements approximately 10 nm in diameter which have been proposed to consist of 5–6 bacteriochlorophyll c protein subunits arranged in a ring-like structure. Between each chlorosome and the cytoplasmic membrane is a crystalline baseplate 5–6 nm thick, which is thought to be a two-dimensional crystalline array of the water-soluble bacteriochlorophyll a protein. At each chlorosome attachment site the cytoplasmic membrane contains 20–45 intramembrane particles which may correspond to complexes, which have been isolated, containing the reaction centres and the water-insoluble Bchl a protein. Each reaction centre complex is apparently directly linked with two trimers of the water-soluble bacteriochlorophyll a protein.

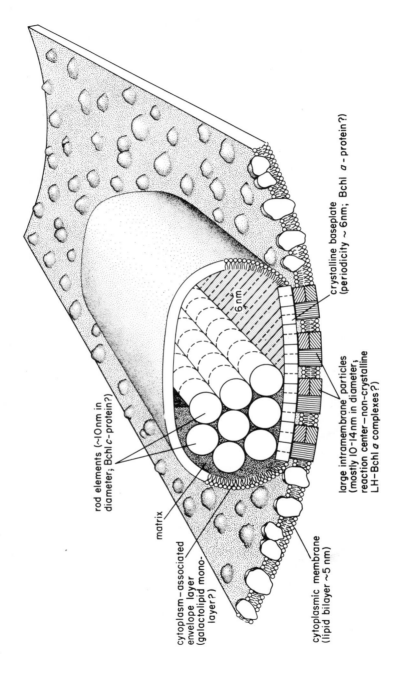

Fig. 3.10 Model of a chlorosome and its associated cytoplasmic membrane based on freeze-fracture studies of *Chlorobium limicola* f. *thiosulfatophilum*. The hydrophobic core of the chlorosome contains 10–30 rod elements surrounded by matrix material. The 2–4 nm wide envelope layer adjacent to the cytoplasm is thought to consist of a layer of galactolipid molecules. A crystalline baseplate, 5–6 nm thick, connects the chlorosome core to about 30 large intramembrane particles inside the cytoplasmic membrane. (Reproduced, with permission, from Staehelin *et al.*, 1980).

Fluorescence studies of the green sulphur bacteria have shown that light energy absorbed by the major light-harvesting pigments bacteriochlorophyll c, d or e is transferred to Bchl a and then to the reaction centres. The arrangement and absorption spectra of the light-harvesting pigments reflect this sequence of excitation energy transfer.

There are only relatively slight differences between the organization of the light-harvesting apparatus in the gliding filamentous bacterium *Chloroflexus aurantiacus* and that of *Chlorobium*. The rod elements in the chlorosomes of *Chloroflexus* are only 5.2 nm in diameter, and are presumably different in structure. In *Chlorobium*, the crystalline baseplate of water-soluble Bchl a protein remains associated with the cytoplasmic membrane when the chlorosomes are separated, but the chlorosomes from *Chloroflexus* tend to retain the crystalline baseplate on separation from the membrane. The reaction centre bacteriochlorophyll of *Chloroflexus* is distinctly different, with absorption peaks at 805 and 865 nm and a bleaching at 865 nm upon oxidation (and is called P_{865}). Olson (1981) has recently written a detailed review of the organization of bacteriochlorophyll in green phototrophic bacteria.

3.3.2 Photosynthetic reaction centres in green bacteria

The first indication that the reaction centres of green sulphur bacteria might differ from those of purple bacteria was provided by studies of the pathway of NAD^+ reduction (Buchanan & Evans, 1969). The primary electron acceptor in the reaction centre of purple bacteria has a mid-point redox potential (E_m) of -100 to -200 mV. This makes it unlikely that these reaction centres can reduce NAD^+ ($E_m = -340$ mV) directly. Instead it has been shown that NAD^+ in these organisms is reduced by reverse electron transfer at the expense of pmf (see Section 4.4) or of ATP phosphorylated by the light-driven cyclic electron transport chain. As a result NAD^+ reduction in purple bacteria is sensitive to uncouplers and inhibitors of this cyclic electron transport chain.

In contrast *Chlorobium* is capable of photoreducing ferredoxin and NAD^+ by a pathway that is insensitive to these uncouplers and inhibitors (Evans, 1969). This indicates a very much lower redox potential for the primary electron acceptor in the reaction centres of these bacteria. In fact studies have shown that P_{840} (the primary electron donor) bleaching is blocked at low redox potentials with a midpoint potential of -550 mV (Jennings & Evans, 1977) suggesting a primary electron acceptor with this potential. This would be capable of reducing NAD^+ directly. Acting as electron donor for the reaction centre *in vivo* are sulphide or thiosulphate, which are oxidized by specific reductases through cytochromes c_{551} and c_{553}. Both of these cytochromes then transfer the electron to cytochrome c_{555} which donates the electron to the reaction centre. The electron is then transferred from the primary acceptor through a soluble ferredoxin and flavoprotein reductase to NAD. The non-cyclic electron transfer through this reaction centre is very similar to that of photosystem I in plants.

The light-induced optical difference spectrum of P_{840} in green sulphur bacteria is very different from P_{870} and P_{960} in purple bacteria. Two narrow peaks (half-width 5 nm) appear at 830 nm and 842 nm, instead of the single broad peak (half-width 25 nm) of P_{870} or P_{960}. These results could indicate that green sulphur bacteria contain two types of reaction centre. The light-induced EPR signal attributed to the primary electron donor in membrane preparations from *C. limicola* is similar to that observed in the reaction centres of purple bacteria. At cryogenic temperatures, however, photo-oxidation of the reaction centres as indicated by this EPR signal is partly irreversible and partly reversible, again suggesting that two types of reaction centre may be present.

EPR studies have shown (Knaff & Malkin, 1976) that cell-free preparations from *C. limicola* contain several low-potential iron-sulphur proteins, which could act as primary electron acceptor. Jennings & Evans (1977) described the irreversible photo-reduction at cryogenic temperatures, of an iron-sulphur centre ($E_m = -550$ mV; Fig. 3.11) which is probably the best candidate for the

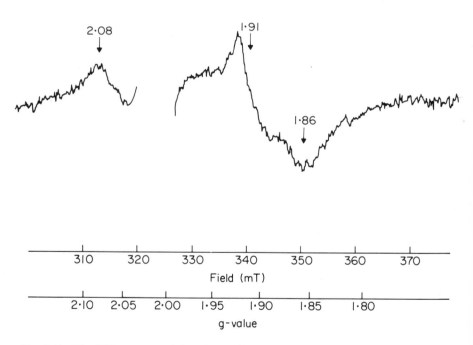

Fig. 3.11 The EPR spectrum of the primary electron acceptor irreversibly photoreduced at cryogenic temperatures in reaction centre preparations from the green sulphur bacterium *Chlorobium limicola* f. *thiosulfatophilum*. The spectrum, the difference between the spectrum of a dark sample and its spectrum following illumination at 15 K, was recorded at 15 K.

primary electron acceptor. The reversible component of the photo-oxidation of P_{840} at cryogenic temperatures is not linked to the reduction of this iron-sulphur centre, again suggesting the presence of two different types of reaction centre.

It is not known whether the green filamentous gliding bacteria are capable of the direct photoreduction of NAD^+. However, the light-induced optical difference spectrum of P_{865} resembles that of P_{870} and P_{960} of purple bacteria, with a single broad band at 865 nm. This primary electron donor also has a typical bacteriochlorophyll a dimer EPR signal upon oxidation, although photo-oxidation is partly reversible and partly irreversible at cryogenic temperatures.

3.4　SUMMARY

The light-harvesting and reaction centre systems of the phototrophic bacteria provide a mechanism for the conversion of light energy to the chemical energy of a reducing agent and a transmembrane potential. They are incorporated into the electron transport chain of the organism to allow production of ATP and NADH (Fig. 3.12).

The photochemical events in the purple bacterial reaction centre initially operate with very high efficiency. The energy available from a quantum of light at 870 nm, the functional wavelength, is 1.4 V. The redox difference of the initial charge separation between P_{870} and the bacteriopheophytin is about 1.05 V. However, a considerable part of this energy is lost in stabilizing the charge separation to the quinone with a final redox difference between the P_{870} and secondary quinone of only 0.5 V. The overall efficiency is therefore only about 35% although some of the energy is also recovered as a membrane potential which may be used for transport and ATP synthesis. The redox potential of the acceptor is about -150 mV so that NADH can only be produced indirectly.

Less is known about primary photochemistry in the green phototrophic bacteria. However, what is known is intriguing. As an adaptation to growth at very low light intensities green sulphur bacteria contain a large number of bacteriochlorophyll c, d or e molecules in addition to the normal complement of bacteriochlorophyll a molecules. These are on the inner surface of the membrane, in the chlorosomes. This is reminiscent of the organization of the phycobilin light-harvesting pigments in phycobilisomes on the surfaces of the photosynthetic membranes in the cyanobacteria. The reaction centre, which has been partially characterized, differs from that of the purple bacteria in having an altered redox range. The reaction centre chlorophyll has a midpoint potential of about $+250$ mV and the electron acceptor -550 mV. This shift allows direct reduction of NAD^+ by a non-cyclic mechanism.

It is interesting to speculate that green sulphur bacteria may contain two

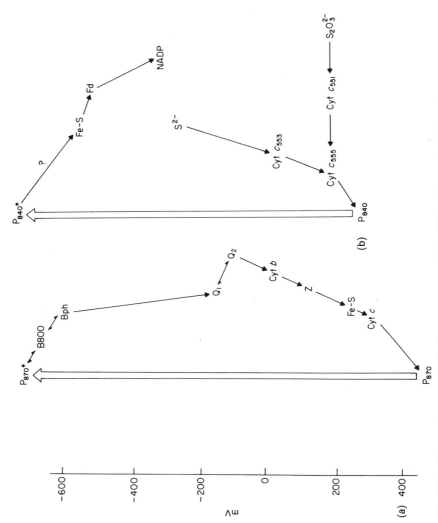

Fig. 3.12 The reaction centre in the electron transport chain. (a) Purple bacteria. (b) Green bacteria. Fe-S: iron-sulphur centre. Fd: soluble ferredoxin. Other abbreviations as in Fig. 3.9.

reaction centres, one resembling photosystem I of cyanobacteria and higher plants, the other possibly like that in purple bacteria, which seems to be similar to photosystem II in plants. This reaction centre might be involved in cyclic electron transport for production of membrane potential and ATP synthesis. The possible presence of two reaction centres, the organization of the chlorosomes and the existence of filamentous gliding forms may indicate that the green bacteria represent a stage in evolution between the purple bacteria and the cyanobacteria.

Acknowledgements

We would like to thank Dr J Olson for sending his review prior to publication.

REFERENCES

BLANKENSHIP R. E. & PARSON W. W. (1978) The photochemical electron transfer reactions of photosynthetic bacteria and plants. *Ann. Rev. Biochem.*, **47**, 635–53.

BOWMAN M. K., NORRIS J. R. & WRAIGHT C. A. (1979) Distance determination in bacterial photoreaction centres by electron spin relaxation *Biophys. J.*, **25**, 203A.

BUCHANAN B. B. & EVANS M. C. W. (1969) Photoreduction of ferredoxin and its use in NAD (P)$^+$ reduction by a subcellular preparation from the photosynthetic bacterium *Chlorobium thiosulfatophilum. Biochem. Biophys. Acta*, **180**, 123–9.

COGDELL R. J. & THORNBER J. P. (1979) The preparation and characterization of different types of light-harvesting pigment-protein complexes from some purple bacteria. In Chlorophyll Organization and Energy transfer in Photosynthesis. (Ed. by G. Wolstenholme & D. W. Fitzsimons), pp. 61–72. Ciba Symposium 61.

COMMONER B., HEISE J. J. & TOWNSEND J. (1956) Light induced paramagnetism in chloroplasts. *Proc. natn. Acad. Sci. USA*, **42**, 710–18.

DUTTON P. L. & PRINCE R. C. (1978) Reaction centre driven cytochrome interaction in electron and proton translocation and energy coupling. In *The Photosynthetic Bacteria* (Ed. by R. K. Clayton & W. R. Sistrom), pp. 525–70. Plenum Press, New York.

DUTTON P. L., PRINCE R. C., TIEDE D. M., PETTY K. M., KAUFMANN K. J., NETZEL T. L. & RENTZEPIS P. M. (1977) Electron transfer in the photosynthetic reaction centre. In Brookhaven Symposia in Biology No. 28 (Ed. by J. M. Olsen & G. Hind), pp. 213–37.

EVANS M. C. W. (1969) Ferredoxin: NAD reductase and the photoreduction of NAD by *Chlorobium thiosulfatophilum*. In *Progress in Photosynthesis Research* (Ed. by H. Metzner) Vol. 3 pp. 1474–5. Loupp, Tübingen.

FAJER J., DAVIS M. S., BRUNE D. C., SPAULDING L. D., BORG D. C. & FORMAN A. (1977) Chlorophyll radicals and primary events. In Brookhaven Symposia in Biology No. 28 (Ed. by J. M. Olson & G. Hind), pp. 74–104.

FEHER G. & OKAMURA M. Y. (1978) Chemical composition and properties of reaction centres. In *The Photosynthetic Bacteria* (Ed. by R. K. Clayton & W. R. Sistrom), pp. 349–86. Plenum Press, New York.

HOPFIELD J. J. (1974) Electron transfer between biological molecules by thermally activated tunneling. *Proc. natn. Acad. Sci. (USA)*, **71**, 3640–4.

JENNINGS J. V. & EVANS M. C. W. (1977) The irreversible photoreduction of a low potential component at low temperatures in a preparation of the green photosynthetic bacterium *Chlorobium thiosulfatophilum. FEBS Lett.*, **75**, 33–36.

KLIMOV V. V., SHUVALOV V. A., KRAKHMALEVA I. N., KLEVANIK A. V. & KRASNOVSKY A. A. (1977) Photoreduction of bacteriopheophytin *b* in the primary light reaction of chromatophores of *Rps. viridis. Biokhimiya*, **42**, 519–30.

KNAFF D. B. & MALKIN R. (1976) Iron-sulphur proteins of the green photosynthetic bacterium *Chlorobium. Biochim. Biophys. Acta*, **430**, 244–52.

LEIGH J. S. & DUTTON P. L. (1972) The primary electron acceptor in photosynthesis. *Biochem. Biophys. Res. Commun.*, **46**, 414–21.

MARINETTI T. D., OKAMURA M. Y. & FEHER G. (1979) Localisation of the primary quinone binding site in reaction centres from *Rps. sphaeroides* by photoaffinity labeling. *Biochemistry*, **18**, 3126–33.

MATTHEWS B. W., FENNA, R. F., BOLOGNESI M. C., SCHMID M. F. & OLSON J. M. (1979) Structure of a bacteriochlorophyll *a* protein from the green photosynthetic bacterium *Prosthecochloris aestuarii. J. Mol. Biol.*, **131**, 259–85.

NORRIS J. R. & KATZ J. J. (1978) Oxidized bacteriochlorophyll as photoproduct. In *The Photosynthetic Bacteria* (Ed. by R. K. Clayton & W. R. Sistrom), pp. 397–418. Plenum Press, New York.

OLSON J. M. (1981) Chlorophyll organisation in green photosynthetic bacteria. *Biochim. Biophys. Acta*, **594**, 33–51.

PIERSON B. K. & CASTENHOLZ R. W. (1974) Studies of pigment and growth in *Chloroflexus aurantiacus*, a phototrophic filamentous bacterium. *Arch. Microbiol.*, **100**, 283–305.

PRINCE R. C., LEIGH J. S. & DUTTON P. L. (1976) Thermodynamic properties of the reaction centre of *Rps. viridis. Biochim. Biophys. Acta*, **440**, 622–36.

REED D. W. & CLAYTON R. K. (1968) Isolation of a reaction center fraction from *Rps. sphaeroides. Biochem. Biophys. Res. Commun.*, **30**, 5, 471–5.

RUTHERFORD A. W. & EVANS M. C. W. (1980) Direct measurement of the redox potential of the primary and secondary quinone electron acceptors in *Rhodopseudomonas sphaeroides* (wild type) by EPR spectrometry. *FEBS Lett.*, **110**, 257–61.

RUTHERFORD A. W., HEATHCOTE P. & EVANS M. C. W. (1979) Electron paramagnetic resonance measurements of the electron transfer components of the reaction centre of *Rps. viridis. Biochem. J.*, **182**, 515–23.

SCHENCK C. C., PARSON W. W., HOLTEN D. & WINDSOR M. W. (1981) Transient states in reaction centres containing reduced bacteriopheophytin. *Biochim. Biophys. Acta*, **635**, 383–92.

SHUVALOV V. A. & PARSON W.W. (1981) Electron transfer reactions between P_{870}, Bchl-800 and bacteriopheophytin in bacterial reaction centers. In *Photosynthesis* III (Ed. by G. A. Koynnoglu) pp. 949–58. Balaban International Science Services, Philadelphia.

STAEHELIN L. A., GOLECKI J. R. & DREWS G. (1980) Supramolecular organisation of chlorosomes (Chlorobium vesicles) and of their membrane attachment sites in *Chlorobium limicola. Biochim. Biophys. Acta*, **589**, 30–45.

WRAIGHT C. A. & CLAYTON R. K. (1973) The absolute quantum efficiency of bacteriochlorophyll photooxidation in reaction centres of *Rps. sphaeroides. Biochim. Biophys. Acta*, **333**, 246–60.

Chapter 4. Production and Dissipation of Membrane Potential; Formation of ATP and Reducing Equivalents

M. C. W. EVANS

4.1 INTRODUCTION

Phototrophic organisms obtain all or part of their energy for growth from light. Energy is required in three different, but interconvertible forms: as ATP, as reducing compounds and as ionic or electrical gradients, the proton motive force (pmf). ATP and reductant are required for the biosynthetic processes of growth. ATP and the pmf are used for substrate and ion transport, sensory processes and movement. The primary events in the conversion of light into chemical energy take place in the photosynthetic reaction centre and are described in detail in Chapter 3. In green sulphur bacteria these events result in the direct formation of a low potential reductant suitable for CO_2 or N_2 fixation. In purple bacteria they do not and alternative mechanisms must be used for this purpose.

In purple bacteria the direct role of the reaction centre is the formation of a reductant of moderate potential which can be oxidized *via* the electron transport chain. Both the primary events in the reaction centre and the secondary electron transport result in generation of a membrane potential and subsequently ATP synthesis. In green bacteria virtually nothing is known about the mechanism of electron transport and ATP synthesis. A small number of papers have been published on ATP synthesis in green sulphur bacteria but the results are unsatisfactory and at present it can only be assumed that they have a system based on the same principles as those in other organisms. This chapter concentrates on the process of membrane potential generation and ATP synthesis in purple bacteria. Even amongst these organisms research has been largely restricted to three or four species, namely *Rhodopseudomonas sphaeroides*, *Rp. capsulata*, *Rhodospirillum rubrum* and to a small extent *Chromatium vinosum* or *Chr. minutissima*. It is assumed that the coupling mechanism between electron transport and ATP synthesis is essentially as described by the chemiosmotic theory, i.e., an indirect coupling. According to this theory, the function of electron transport is the generation of a 'proton motive' force which is subsequently used for ATP synthesis. The latter process depends upon the activity of a membrane bound enzyme, the ATP synthase.

4.2 THE ELECTRON TRANSPORT CHAIN

The electron transport chain in purple bacteria can be conveniently considered to have two major components. One is the reaction centre and the other a cytochrome bc complex. The bc complex is very similar to the central section of the mitochondrial electron transport chain, complex III, and in facultatively aerobic species of purple non-sulphur bacteria, the cytochrome complex is identical in the photosynthetic and respiratory chains. The reaction centre functions as both oxidant and reductant for the cytochrome complex, accepting electrons from Cyt c and donating electrons through ubiquinone to Cyt b, resulting in a light activated cyclic electron transport system. The pathway of electron transfer in the complex is still not clearly understood despite many years of intensive study. Most information has come from studies of the kinetics of oxidation and reduction of the cytochromes using dual wavelength spectrophotometry. In most experiments measurements are made of absorption changes following a short flash of light which causes a single turnover of the reaction centre, i.e. it makes one electron travel round the electron transport system. The effect of inhibitors, and changes in pH or ambient redox potential have also been investigated. These experiments have become extremely complex since the development of computer controlled spectrophotometers which offer automatic control of many environmental parameters. Only an outline of these experiments can be given in this chapter; detailed discussions are available in reviews by Prince *et al.*, (1982) and Crofts & Wraight (1983) and the major book edited by Clayton & Sistrom (1978).

4.2.1 Components of the bc complex

Cytochrome c

Two c type cytochromes have been identified in the electron transport chain of purple bacteria (Woods, 1980). One of these, Cyt c_2, was for many years thought to be the only c type cytochrome present. It is a soluble protein situated on the periplasmic side of the membrane. It can be functionally inhibited by antibodies or removed by washing of spheroplasts (Prince *et al.*, 1975). The second, recently discovered Cyt c_1 is a membrane bound protein. The two cytochromes cannot be easily distinguished spectrophotometrically and appear to have essentially the same redox potential. Other c type cytochromes in green and purple sulphur bacteria are involved in substrate oxidation (Chapter 5).

Cytochrome b

Three b type cytochromes can be distinguished in *Rp. sphaeroides* (Dutton & Jackson, 1972). These are b_{561}, b_{560} and b_{566}. They are most easily distinguished by their midpoint potentials at pH 7.0: $b_{561} = +155$, $b_{560} = +50$ and $b_{566} = -90$ mV. A similar group of b type cytochromes have been characterized in *Rp. capsulata* (Evans & Crofts, 1974) and b type cytochromes have been identified in green and purple sulphur bacteria (Knaff & Buchanan, 1975).

Iron-sulphur centres

A number of iron-sulphur centres can be identified in chromatophore membranes (see Evans, 1982). These include the centres associated with succinic dehydrogenase and NAD dehydrogenase. Only one centre is directly involved in cyclic electron transport. This is the 'Rieske' iron-sulphur centre which is a two-iron two-sulphur centre with $Em_{7.0} = +280$ mV. It is unusual in that its redox potential is pH dependent, which may reflect close association with a quinone. It can only be detected by EPR spectrometry at low temperatures (20 K). Its kinetic properties can therefore only be determined indirectly from the behaviour of the cytochromes which react with it.

Quinones

The facultatively phototrophic purple bacteria contain ubiquinone. Some at least of the strictly phototrophic bacteria such as *Rp. viridis* and a *Chromatium sp.* have in addition menaquinone in their reaction centres (Cogdell, 1982).

The quinones have multiple functions in the electron transport chain. Two quinone molecules, either ubiquinone or menaquinone depending on the bacterial species, are bound to the reaction centre (see Section 3.2.6). A large pool of ubiquinone (20–30 per Cyt c) is also present. Quinone extraction experiments suggest that this pool is required for ATP synthesis or continuous electron transport but not for electron transport following a single turnover flash. However, a third quinone molecule is apparently essential for electron transport. Originally identified kinetically as the reductant for Cyt c, extraction experiments identify it as a quinone Qz (Takamiya & Dutton, 1979). It is not yet clear whether Qz is protein bound as are the reaction centre quinones or is an essential 'mobile' quinone.

Investigations of electron transport through quinones is almost as difficult as for iron-sulphur proteins. When a relatively stable semiquinone is involved as with the reaction centre quinones, changes can be detected both optically and by EPR. However, when, as seems to be the case with Qz or redox changes of the quinone pool, two electrons are transferred simultaneously the only absorption changes are in the ultraviolet region where it is technically almost impossible to make useful kinetic measurements. There is therefore no direct evidence for the existence of Qz.

The Cyt bc complex has been isolated from mitochondria and recently from phototrophic organisms including *Rp. sphaeroides* (Gabellini et al., 1982).

The mitochondrial complex contains perhaps nine subunits. The complex from *Rp. sphaeroides* is simpler and contains only three polypeptides of 40, 34, and 25 KD. The 34 KD peptide is identical with Cyt c_1 and the preparation contains 1 c_1, 2 b type haem, 1 Fe-S centre and 0.4–0.6 ubiquinone. It is active in electron transport from ubiquinol to Cyt c_2 and its activity is inhibited by antimycin and 5-n-undecyl-6-hydroxy-4,7-dioxobenzothiazole (UHDBT).

4.2.2 The pathway of electron transfer

It is clear from the current literature that a definitive description of the pathway of electron transfer in the cyclic electron transport system of purple bacteria cannot be made at present. Classically, conceptions of electron transport chains have been based on the redox and kinetic properties of the

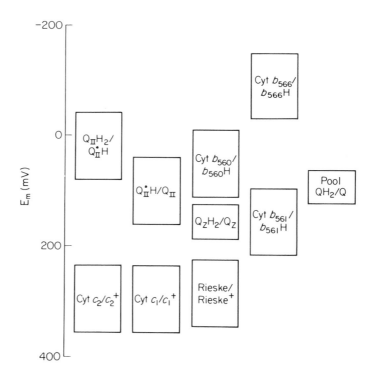

Fig. 4.1 Thermodynamic profile of the components of the ubiquinone-cytochrome c_2 oxidoreductase of *Rps. sphaeroides*. The values reflect those measured at pH 7, and the boxes enclose the E_h range over which a component changes from being 91 % oxidized to 91 % reduced. The E_m values of cytochromes c_2 and c_1 have not yet been unambiguously separated, but they do appear approximately equipotential. The values for QII are extrapolated from the data of Rutherford & Evans (1980). Reproduced from Prince *et al.*, 1982, with permission.

components. However, as can be seen in Fig. 4.1, there is considerable overlap of potentials among the components of the chain. Kinetic measurements provide a fairly clear picture in the cytochrome c region but not for the two b type cytochromes which seem to be involved, b_{560} and b_{561}. The situation is further complicated by the fact that the second quinone of the reaction centre appears to act as a 'gate' (see Section 3.2.6), donating electrons to the b cytochromes two at a time (Wraight, 1980). It also seems likely that there are more reaction centres than cytochrome complexes, perhaps in a two to one ratio. The fact that virtually all available information is based on measurements following a single turnover flash rather than continuous illumination may also complicate the interpretation of the normal operation of the system. The simplified model of the electron transport system described below is undoubtedly 'incorrect', but at present there is no 'correct' model available. Fig. 4.2 shows a simplified model of the electron transport chain. Parts of it are well-established, like the sequence of carriers in the reaction centre (see Chapter 3) and the Rieske centre Cyt c section of the chain. Cyt c oxidation following flash illumination is biphasic, with one component oxidized in 3 μs the other in 100 μs or slower. These two phases presumably represent the oxidation of c_2 and c_1 respectively. In the presence of the inhibitor UHDBT which prevents electron transfer from the Rieske centre to c_1, a saturating flash, oxidizing all the reaction centres, causes the oxidation of only one mole

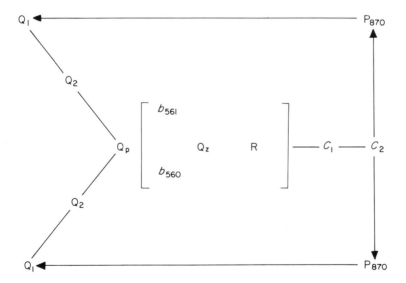

Fig. 4.2 The electron transport chain of *Rp. sphaeroides*. Q_1, Q_2: bound quinones in the reaction centre. Q_p: the quinone pool. Q_z: quinone components involved in reduction of the Rieske centre. b: b type cytochrome, c: c type cytochrome. R: Rieske iron-sulphur centre, P_{870}: the reaction centre chlorophyll.

of Cyt c per reaction centre, i.e. $\frac{1}{2} c_1$, and $\frac{1}{2} c_2$ per reaction centre. Assuming there are two reaction centres per cytochrome chain, each of a pair of reaction centres must accept an electron, the c_2 acting as a mobile carrier between the bound c_1 and the two reaction centres. In the absence of UHDBT the Rieske iron-sulphur centre equilibrates rapidly with the c cytochromes, so it is not possible to tell if it is oxidized by c_1 or c_2. The Rieske centre is reduced by the quinone Qz. This reduction is sensitive to the inhibitor antimycin A, although it is not the site of antimycin inhibition. Initially, Qz was proposed as a carrier required to complete the chain as defined by kinetic experiments. Subsequently, extraction experiments showed it to be a quinone. The major difficulties in understanding the electron transport chain are in the Cyt b region. There is good evidence for the involvement of b_{560}. When it is reduced in the dark, flash illumination causes transient oxidation, and when it is oxidized in the dark, a flash causes reduction. At high redox potentials when Qz is oxidized, cytochrome b_{560} remains reduced for some time. When Qz is initially reduced, however, the b_{560} becomes rapidly reoxidized. There is recent evidence that b_{566}, the low potential b type cytochrome, is also involved in cyclic electron flow.

The difficulties in explaining electron transfer in the b cytochrome region of the chain arise in part from the fact that this section of the chain, as in mitochondria, functions as a 'Mitchell loop' transferring protons from one side of the membrane to the other, which any model must explain, and partly from problems in interpretation of experimental results. These include the role of b_{566}, the inhibition by antimycin of the oxidation of Qz by Cyt c (antimycin almost certainly reacts close to one of the b cytochromes), and the odd phenomenon of oxidant-induced reduction of Cyt b. This latter effect is observed when a small amount of ferricyanide is added to chromatophores with Qz reduced in the presence of antimycin. A transient oxidation of Cyt c and reduction of Cyt b is observed. The best explanation for this phenomenon, and consequently a candidate for the normal mechanism of electron transfer, is a Q cycle of the type proposed by Mitchell (1975) (see also Nicholls, 1982), Fig. 4.3. In this scheme illumination results in oxidation of the reaction centre and subsequently of the c cytochromes and the Rieske centre which is then reduced by a single electron from Qz, leaving Qz as the low potential semiquinone which in turn reduces the b cytochrome. To complete the cycle the electron from the b cytochrome must be transferred back into the electron flow. According to a classical Q cycle, one electron from the b cytochromes and one from the reaction centre would be transferred to Qz. However, as it is clear that the reaction centre donates electrons to Qz (or a pool quinone) two at a time, quinone reduction by the b cytochromes must also involve transfer of two electrons to a quinone. This type of scheme requires three reaction sites for mobile (or pool) quinone: a reductive site on the reaction centre and both oxidative and reductive sites on the bc complex. Quinones reduced either at the reaction centre or at the reductive site of the bc complex would react identically

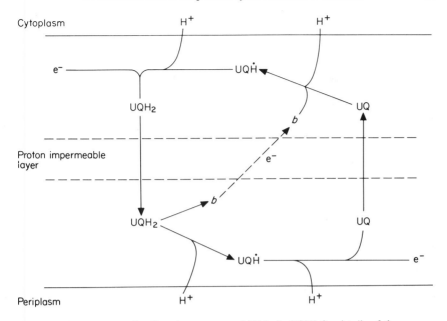

Fig. 4.3 The proton motive Q cycle, see text and Nicholls (1982) for details of the operation of the cycle.

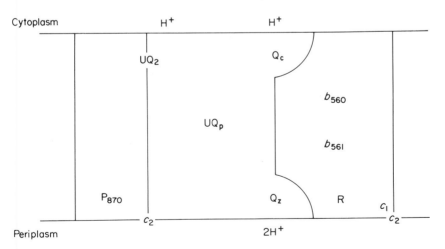

Fig. 4.4 The electron transport chain in the cell membrane of purple bacteria. See Fig. 4.2 for abbreviations. Q_z: quinone oxidizing site of the *bc* complex. Q_c: quinone reducing site of the *bc* complex. UQ_2: bound ubiquinone of the reaction centre. UQ_p: ubiquinone pool.

at the oxidative site of the *bc* complex. A single quinone would clearly be unable to sustain electron flow through such a complex system and the size of the quinone pool probably reflects the optimum quinone concentration for electron flow under continuous illumination. The overall effect is that each electron is cycled through the *b* cytochrome region twice.

Photosynthesis in purple bacteria depends on electron transport from the quinone acceptors of the reaction centre through *b* and *c* type cytochromes, quinones and the Rieske centre to the oxidized reaction centre chlorophyll. These components are spatially organized and Fig. 4.4 shows a simple model describing this.

4.3 ATP SYNTHESIS

The synthesis of ATP is probably the major function of the photosynthetic apparatus in purple bacteria under heterotrophic growth conditions. Shortly after the discovery of photophosphorylation in chloroplasts Frenkel (1954) showed that chromatophores of purple bacteria were able to catalyse the synthesis of ATP.

$$ADP + P_1 \xrightarrow{\text{Light}} ATP$$

This ATP synthesis is coupled to the cyclic electron transport chain discussed above. Inhibition of electron transport prevents ATP synthesis. Synthesis of ATP is also inhibited by classical uncouplers of oxidative phosphorylation and by appropriate combinations of ionophores.

That coupling of the light-driven reaction to ATP synthesis is indirect can be readily shown by two types of experiment. Synthesis of ATP can be measured in response to flashes of light; ideally each flash should give a single turnover of the reaction centre. In experiments of this type it is found that the optimum ATP yield per flash is observed if there is a relatively long dark interval, at least 20 ms, between flashes (Nishimura, 1962). Since the turnover time of the uncoupled electron transport chain is 1–2 ms, slow dark reactions must be involved. An ATP synthesis rate of about 1 ATP per reaction centre every 100 ms has been estimated (Petty & Jackson, 1979). A different type of experiment also indicates the existence of a long-lived dark intermediate. If chromatophores are illuminated in the absence of ADP and then injected into an ADP solution in the dark, small amounts of ATP are synthesized. These experiments are more difficult to do than similar ones in chloroplasts, but it can clearly be shown that electron transport is only required in the light (Leiser & Gromet-Elhanan, 1974).

In the absence of evidence for any stable chemical intermediates the chemiosmotic hypothesis offers the best model for the nature of the intermediate as a proton motive force, consisting of a mixture of proton (pH) gradient and membrane potential, which is used to drive an anisotropic proton ATPase. The purple bacteria have proved to be very useful organisms for the investigation of the formation of the pmf. The steps in energy conversion are summarised in Fig. 4.5.

4.3.1 Proton gradient formation

Direct observations of proton pumping can be made using a pH electrode (Scholes *et al.*, 1969). Protons are pumped on illumination from the cytoplasmic to the periplasmic side of the membrane, resulting in acidification of the medium if whole cells are used or alkalinization when chromatophores are used. Formation of a pH gradient is prevented by electron transport inhibition and the gradient is dissipated by uncouplers. It is not possible to measure the true kinetics of formation of the gradient with a pH electrode, or, since there is no net electron transport, to estimate the H^+/e ratio. This difficulty can be overcome by following rapid pH changes using pH indicating dyes such as cresol red and measuring changes in absorption by dual wavelength spectrophotometry. Experiments of this type show that following a single turnover flash, two H^+ per reaction centre are bound by the chromatophore membrane. The binding of one of these protons is inhibited by antimycin, showing that one proton is bound by the reaction centre and one by the *bc* complex. The binding of one proton can also be observed in isolated reaction centres. A value of two protons per electron transport chain can only be observed in the presence of valinomycin, an ionophore which allows equilibration of K^+ across the membrane, enabling complete expression of the pmf as a pH gradient.

The simplest explanation for the binding of protons to the cytoplasmic side of the membrane and their subsequent release on the periplasmic side would be that originally proposed by Mitchell, that is that the proton is bound in the reduction of the quinones (Q_2 and Qz) which then cross the membrane and are reoxidized on the other side releasing the proton. However, data on the pattern of H^+ binding and the pK values for H^+ binding compared to the pK values of the quinones suggest that this may be too simple and that protein binding sites and Cyt b_{560} may be involved. The mechanism of proton movement across the membrane may require 'mobile' quinone as the quinone pool is required for ATP synthesis in continuous light. If a Q cycle type of mechanism is operating, two H^+ should be bound for each electron passing through the *bc* complex. This is apparently not the case, as shown by the results of experiments on H^+ binding. However, if there are two reaction centres per complex, the equivalence of H^+ bound by the reaction centre and the *bc* complex may actually be the result of two reaction centres binding one H^+ and one *bc* complex binding two. If there are two reaction centres per complex, the use of saturating flashes may be misleading, as under the light intensities usually encountered during growth it may be that only one of the reaction centres is likely to be activated at any time. Under saturating flash conditions the excitation of both reaction centres may therefore either result in anomalous electron flow with electrons available from both centres simultaneously, or one centre may be unable to pass on the electron and undergo a back reaction from Q_2 to P_{870}. The situation may be even more complex in *Rs. rubrum*, which may have six reaction centres per *bc* complex. The ratio may also vary in response to light intensity during growth.

4.3.2 Membrane potential generation

The development of the pH component of the pmf depends on the transport of protons. This is achieved in effect by transporting hydrogen atoms in one direction across the membrane and electrons in the reverse direction. This electron transfer, if not accompanied by proton transfer and accumulation, gives rise to a membrane potential, i.e. a charge differential across the membrane. The membrane potential ($\Delta\Psi$) can be very simply measured in chromatophores. Some of the carotenoids associated with the reaction centre undergo an electrochromic shift of their absorption maxima when a potential is applied across the chromatophore membrane (Jackson & Crofts, 1969). This shift is observed as an absorption change, the carotenoid band shift, in the 520 nm region and can be followed spectrophotometrically in the same way as, for example, changes in the oxidation state of cytochromes. Measurements of changes in the carotenoid band shift following either flash illumination or the onset of continuous illumination show fast and slow phases. The fast phase has components in the nanosecond time range associated with charge separation in the reaction centre and in the microsecond time range associated with electron transfer from Cyt c_2 to the reaction centre. The slow phase in the micro to millisecond time range is associated with electron transfer in the *bc* complex. The size of the membrane potential can be estimated by calibrating the carotenoid band shift with a potassium potential in the presence of the K^+ ionophore valinomycin. However, there are differences between measurements of $\Delta\Psi$ by the carotenoid band shift which indicate $\Delta\Psi \simeq 200 - 250$ mV, and those obtained by equilibration of permeant ions such as thiocyanate across the membrane which give values around 150 mV (see Baltscheffsky *et al.*, 1982). The difference is probably due to the fact that only the bulk potential is measured by the permeant ions while bulk potential plus the localized charge separation potential in the reaction centre are measured by the carotenoid band shift. In *Rp. sphaeroides* the Δ pH at pH 7.5 has been measured by NMR techniques (Hellingwerf *et al.*, 1981) and shown to be 1.6 pH units $\equiv$ 96 mV. The value of $\Delta^\mu H^+$ is then 150 mV $+ 96$ mV $= 246$ mV. The relationship between $\Delta\Psi$ and Δ pH is complicated and depends upon the capacitances (in turn related to cell volume and area of membrane), the buffering capacity, the permeability of the membrane to ions and the dielectric constant of the membrane. However, as either $\Delta\Psi$ or Δ pH can be used for ATP synthesis in *Rp. sphaeroides*, variations in the proportions are presumably unimportant.

4.3.3 Conversion of $\Delta^\mu H^+$ into ATP

That the pmf is required for ATP synthesis is clearly shown by the action of uncouplers such as carbonylcyanide m-chlorophenylhydrazone (CCCP) which dissipates the total pmf. However, it would seem that either Δ pH or $\Delta\Psi$ can drive ATP synthesis. Experiments with specific ionophores can be used to dissipate specifically one or other component of the pmf. Nigericin catalyses an exchange of K^+ for H^+ across the membrane, neutralizing Δ pH but maintaining $\Delta\Psi$. Valinomycin allows equilibration of K^+, converting $\Delta\Psi$ to Δ pH. Valinomycin causes only a 20% decrease in the rate of photophosphorylation and nigericin is not inhibitory. A combination of the two antibiotics completely dissipates the pmf and inhibits ATP synthesis (Jackson et al., 1968). However, under flash illumination it appears that only $\Delta\Psi$ is involved in ATP synthesis as under these conditions valinomycin alone is inhibitory (Saphon et al., 1975).

The synthesis of ATP depends on the presence in the membrane of a proton translocating ATP synthase. As is the case in all pmf-linked phosphorylation systems, this ATP synthase has two major components. The F_1 component is a large (ca. 300 KD) protein located on the cytoplasmic (alkaline) side of the membrane. The second component of the system is an integral proteolipid (Fo) which acts as a protonophore and binding site for the F_1 ATP synthase. The F_1 component has been isolated from *Rp. capsulata*, *Rs. rubrum* and *Chr. vinosum*. All have rather similar subunit composition and when isolated show Ca^{2+}-dependent ATPase activity. They act as coupling factors reactivating ATP synthesis in depleted membranes. An intact preparation containing both F_1 and F_0 has also been isolated and shown to be competent in ATP synthesis when reconstituted in phospholipid vesicles in which $\Delta^\mu H^+$ is generated by purple membrane from *Halobacterium halobium* (Oren et al., 1980). The mechanism of ATP synthesis by this enzyme and the similar ones from other organisms is not yet understood.

4.4 LOW POTENTIAL REDUCTANT GENERATION

Purple bacteria require NADH (-340 mV) and reduced ferredoxin [-490 mV (Tagawa & Arnon, 1968) or $= -550$ mV (unpublished results from my laboratory)] for CO_2 fixation, N_2 fixation and other biosynthetic processes. Although NADH may be available from substrate oxidation during heterotrophic growth, growth with thiosulphate as electron donor, or carbon substrates such as succinate, requires that light energy should be used for NAD and ferredoxin reduction. In green sulphur bacteria, as in oxygenic phototrophs, the requirement for NADH and reduced ferredoxin is met by a direct transfer of electrons through the reaction centre in non-cyclic electron transport. In the purple bacteria this does not occur, there being no good evidence for the existence of a photochemical system which directly generates an accessible low potential electron acceptor. However, light driven reduction of NADH can be observed in chromatophores of *Rs. rubrum* with succinate as electron donor (Frenkel, 1958) and also, using fluorescence detection of NADH, in whole cells of *Rp. sphaeroides* (Jones & Whale, 1970). In both cases NAD reduction is prevented by inhibitors of cyclic electron transport and uncouplers (Keister & Yike, 1962). The mechanism of NAD reduction depends on the use of $\Delta^\mu H^+$ to drive electrons 'uphill' through a membrane-bound NAD dehydrogenase. This reverse electron transport occurs widely in chemoautotrophic bacteria and can also be demonstrated experimentally in mitochondria. The same mechanism appears to be used also by *Rp. capsulata* growing on H_2, electrons entering the system through a ubiquinone FAD linked hydrogenase, although direct reduction of NAD is theoretically possible (Klemme, 1969; see also Section 5.2.3).

The mechanism of electron transfer from high potential donors such as thiosulphate is not known. Electrons from these enter the electron transport chain through c type cytochromes (Chapter 5) and initial electron transfer to the ubiquinone level may be by a direct, non-cyclic transfer through the reaction centre. Some purple bacteria also use a very low potential ferredoxin for the reductive carboxylation of acetate (see Chapter 6). Reduction of this ferredoxin by reverse electron transport would require an additional energy-dependent step beyond NAD; again no evidence as to how this is achieved is available. The same applies to reduced ferredoxin required for dinitrogen fixation (Chapter 7).

4.5 ORGANIZATION OF THE CHROMATOPHORE MEMBRANE

The existence of specific pathways of electron transport and the coupling of electron transport to pmf generation imply a high degree of organization. The chemiosmotic hypothesis requires the generation of a stable transmembrane pmf and its use by a transmembrane ATP synthase, and, in the wider area of ion transport and motility, other transmembrane enzymes. This implies specific membrane organization and a sealed membrane system. In an intact organism the whole cell acts as the sealed system, while breaking the cells releases membrane fragments which self-seal to form chromatophores. The great majority of these are inside out compared to whole cells, having the cytoplasmic side of the membrane exposed to the medium. There is a large amount of evidence for specific organization of protein in membranes and of specific orientations of prosthetic groups such as haems, chlorophylls and quinones to the membrane plane. This information comes from a number of different types of experiment. Direct examination by electron microscopy can reveal components such as the ATPase. Removal of peripheral proteins from preparations of known sidedness can, for example, identify cytochrome c_2 as a periplasmic protein. Inhibition or agglutination by antibodies to isolated components or chemical labelling of polypeptides in preparations of known sidedness can identify the side on which the membrane proteins are exposed. Optical and EPR studies of oriented membranes provide evidence on the

1. BChl photooxidation. Charge separation at the reaction centre detected by carotenoid bandshift $10^{-12} - 10^{-8}$ s

2. Electron transfer from cyt c to P_{870} and from the quinone acceptors to quinone pool. Proton binding to reaction centre $10^{-6} - 10^{-4}$ s

3. Electron and hydrogen transfer through the bc complex. Slow phase of carotenoid bandshift. Binding of proton on cytoplasmic side of membrane, release on the periplasmic side $10^{-3} - 10^{-2}$ s

4. Accumulation of pmf with repetitive cycling of 1–3

5. Proton influx through ATPase resulting in ATP synthesis or through NAD dehydrogenase resulting in NAD reduction by reverse electron flow.

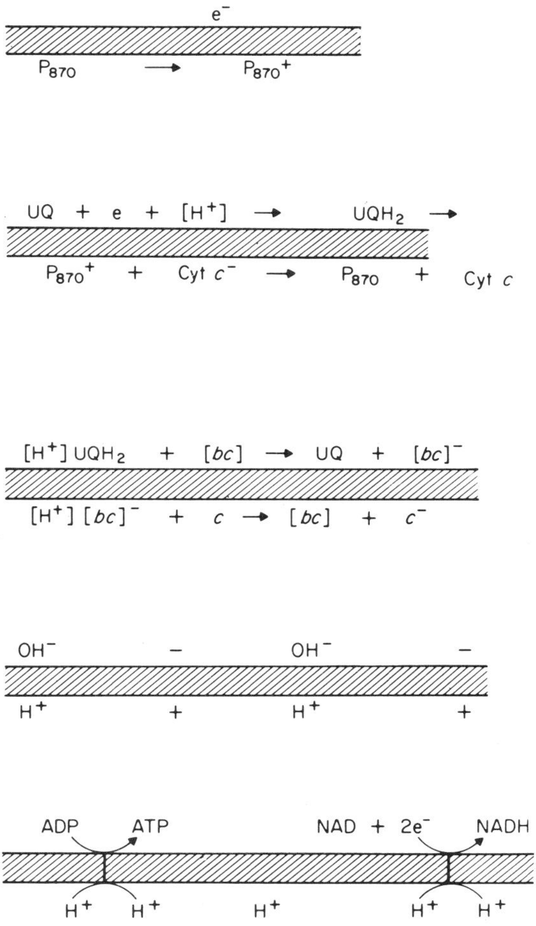

Fig. 4.5 The sequence of events in membrane potential generation and ATP synthesis in membranes of purple bacteria.

orientation of prosthetic groups, and studies of magnetic interactions between groups can provide information on distances between them.

The reaction centre is oriented across the membrane (see Chapter 3 for evidence). This orientation is functional as well as physical, allowing generation of $\Delta\Psi$ in the primary charge separation.

Unfortunately rather little is known about the orientation of the other components of the electron transport chain. Cyt c_2 is a periplasmic protein (Prince et al., 1975) and since it reacts with cytochrome c_1 and possibly the Rieske centre these must also be situated on that side of the membrane. This together with the ability of the complex to generate $\Delta\Psi$ suggests that a hydrogen carrier (or electron carrier plus H^+ extrusion system) must span the membrane from Q_2 to the periplasmic side. An electron carrier must return the electron to the cytoplasmic side and another hydrogen carrier from there to the Rieske centre. These carriers must be the b cytochromes and Qz respectively but there is no direct evidence to support this statement. Fig. 4.4 shows a final model of the electron transport chain placing the carriers in the membrane. Studies of the orientation of pigments, haems and iron-sulphur centres in membranes indicate that while the large complexes may be laterally mobile in a fluid membrane, the active centres of electron transport proteins are tightly bound within the complexes to provide the highly organized electron and proton transport required for the efficient operation of photosynthetic ATP synthesis.

REFERENCES

BALTSCHEFFSKY M., BALTSCHEFFSKY H. & BOORK J. (1982) Evolutionary and mechanistic aspects of coupling and phosphorylation in photosynthetic bacteria. In *Electron Transport and Phosphorylation* (Ed. by J. Barber) pp. 250–68. Elsevier Biomedical Press, Amsterdam.

CLAYTON R. K. & SISTROM W. R. (Eds) (1978) *The Photosynthetic Bacteria*. p. 946. Plenum Press, New York.

COGDELL R. J. (1982) The electron transport components in photosynthetic bacteria. In *Electron Transport and Phosphorylation* (Ed. by J. Barber) pp. 178–94. Elsevier Biomedical Press, Amsterdam.

CROFTS A. R. & WRAIGHT C. (1983) *Ann. Rev. Biophys. Bioeng.*, (In preparation).

DUTTON P. L. & JACKSON J. B. (1972) Thermodynamic and kinetic characterisation of electron transfer components in situ in *Rps. sphaeroides* and *R. rubrum*. *Eur. J. Biochem.*, **30**, 495–510.

DUTTON P. L. & PRINCE R. C. (1978) Reaction centre driven cytochrome interaction in electron and proton translocation and energy coupling. In *The Photosynthetic Bacteria* (Ed. by R. K. Clayton & W. R. Sistrom) pp. 525–70. Plenum Press, New York.

EVANS E. H. & CROFTS A. R. (1974) A thermodynamic characterisation of the cytochromes of chromatophores from *Rps. capsulata*. *Biochim. Biophys. Acta*, **357**, 78–88.

EVANS M. C. W. (1982) Iron-sulfur centers in photosynthetic electron transport. In *Iron-sulfur Proteins* Vol. 4 (Ed. by T. G. Spiro) pp. 250–84. John Wiley and Sons Inc. New York.

FRENKEL A. W. (1954) Light induced phosphorylation by cell-free preparations of photosynthetic bacteria. *J. Am. Chem. Soc.*, **76**, 5568.

FRENKEL A. W. (1958) Light induced reactions of chromatophores of *R. rubrum*. In Brookhaven Symposia in Quantitative Biology, No. 11, p. 276.

GABELLINI N., BOWYER J. R., HURT E., MELANDRI B. A. & HAUSKA G. (1982) A cytochrome b/c complex with ubiquinone-cytochrome c_2 oxidoreductase activity from *Rdp. spheroides*. *Eur. J. Biochem.*, **126**, 105–11.
JONES O. T. G. & WHALE F. R. (1970) The oxidation and reduction of pyridine nucleotides by *Rps. sphaeroides* and *Chl. limicola f. thiosulfatophilum*. *Arch. Microbiol.*, **72**, 48–59.
KEISTER D. L. & YIKE N. J. (1967) Energy linked reactions in photosynthetic bacteria. I. Succinate-linked ATP-driven NAD^+ reduction by *R. rubrum* chromatophores. *Arch. Biochem. Biophys.*, **121**, 415–22.
KLEMME J. H. (1969) Studies on the mechanism of NAD-photoreduction by chromatophores of the facultative phototroph, *Rhodopseudomonas capsulata*. *Z. Naturforsch.*, **24b**, 67–76.
KNAFF D. B. & BUCHANAN B. B. (1975) Cytochrome b and photosynthetic sulfur bacteria. *Biochim. Biophys. Acta*, **376**, 549–60.
LEISER M. & GROMET-ELHANAN Z. (1974) Demonstration of acid-base phosphorylation in chromatophores in the presence of a K^+ diffusion potential. *FEBS Lett.*, **43**, 267–70.
MITCHELL P. (1975) The proton motive Q cycle: a general formulation. *FEBS Lett.*, **59**, 137–9.
HELLINGWERF K. J., KONINGS W. N., NICOLAY K. & KAPTEIN R. (1981) The light-induced pH gradient in chromatophores of *Rps. sphaeroides* as visualised by [31P] NMR. *Photobiochem. Photobiophys.*, **2**, 311–19.
JACKSON J. B., CROFTS A. R. & VON STEDINGK L. (1968) Ion transport induced by light and antibiotics in chromatophores of *R. rubrum*. *Eur. J. Biochem.*, **6**, 41–64.
JACKSON J. B. & CROFTS A. R. (1969) The high energy state in chromatophores from *Rps. sphaeroides*. *FEBS Lett.*, **4**, 185–9.
NICHOLLS D. G. (1982) *Bioenergetics: an Introduction to the Chemiosmotic Theory*. Academic Press, London.
NISHIMURA M. (1962) Studies on bacterial photophosphorylation. I. Kinetics of phosphorylation in *R. rubrum* chromatophores by flashing light. *Biochim. Biophys. Acta*, **57**, 88–95.
OREN R., WEISS S., GARTY H., CAPLAN S. R. & GROMET-ELHANAN Z. (1980) ATP synthesis catalyzed by the ATPase complex from *R. rubrum* reconstituted into phospholipid vesicles together with bacteriorhodopsin. *Arch. Biochem. Biophys.*, **205**, 503–9.
PETTY K. M. & JACKSON J. B. (1979) Kinetic factors limiting the synthesis of ATP by chromatophores exposed to short flash excitation. *Biochim. Biophys. Acta*, **547**, 474–83.
PRINCE R. C., BACCARINI-MELANDRI A., HAUSKA G. A., MELANDRI B. A., & CROFTS A. R. (1975) Asymmetry of an energy transducing membrane. The location of cytochrome c_2 in *Rps. sphaeroides* and *Rps. capsulata*. *Biochim. Biophys. Acta*, **387**, 212–27.
PRINCE R. C., O'KEFFE D. P. & DUTTON P. L. (1982) The organisation of the cyclic electron transfer system in photosynthetic bacterial membranes. In *Electron Transport and Photophosphorylation*. (Ed. by J. Barber) pp. 198–243. Elsevier Biomedical Press, Amsterdam.
RUTHERFORD A. W. & EVANS M. C. W. (1980) Direct measurement of the redox potential of the primary and secondary quinone electron acceptors in *Rhodopseudomonas sphaeroides* (wild type) by EPR spectrometry. *FEBS Lett.*, **110**, 257–61.
SAPHON S., JACKSON J. B., LERBS V. & WITT H. T. (1975) The functional unit of electrical events and phosphorylation in chromatophores from *Rhodopseudomonas sphaeroides*. *Biochim. Biophys. Acta*, **408**, 58–66.
SCHOLES P., MITCHELL P. & MOYLE J. (1969) The polarity of proton translocation in some photosynthetic micro-organisms. *Eur. J. Biochem.*, **8**, 450–4.
TAGAWA K. & ARNON D. I. (1968) Oxidation reduction potentials and stoichiometry of electron transfer in ferredoxins. *Biochim. Biophys. Acta*, **153**, 602–13.
TAKAMIYA K. & DUTTON P. L. (1979) Ubiquinone in *Rps. sphaeroides*: some thermodynamic properties. *Biochim. Biophys. Acta*, **546**, 1–16.
WRAIGHT C. A. (1980) The role of quinones in bacterial photosynthesis. *Photochem. Photobiol.*, **30**, 767–76.

Chapter 5. Electron Donor Metabolism in Phototrophic Bacteria

T. A. HANSEN

5.1 INTRODUCTION

Electron donors can have two functions in bacterial metabolism: (i) reduction of an oxidant such as O_2, NO_3^- and SO_4^{2-} for energy generation and (ii) reduction of components from the growth medium especially CO_2, for the synthesis of cell material. In the phototrophic bacteria, the latter process is the more important and will be dealt with at length in this chapter. Recently, however, chemolithotrophy was discovered as an additional mode of growth of certain purple bacteria and, therefore, this process is also discussed (see Section 5.3).

The importance of electron donors in the growth of purple bacteria depends on the carbon source. A *Rhodopseudomonas* strain growing phototrophically on a complex medium usually does not need an electron donor. During growth with butyric acid (or longer fatty acids) and CO_2, however, the fatty acid is not only used as a carbon source but also as an electron donor for the fixation of CO_2. One could also describe the role of CO_2-fixation as a means of enabling the organism to grow on a reduced organic substrate. For a discussion of the metabolism of organic carbon sources see Chapter 6. Here, we only deal with true electron donors, namely reduced inorganic sulphur compounds, hydrogen, methanol and formate.

Less than a decade ago anaerobic life in the light, that is with light as the energy source, was thought to be restricted to the purple and green bacteria. Since the discovery of anoxygenic photosynthesis among the cyanobacteria (Cohen *et al.*, 1975a) several papers on this topic have appeared. Data from these studies are included in this chapter (see Section 5.4).

Table 5.1 gives a summary of data on the utilization of electron donors by the various phototrophic prokaryotes. It is an old but often forgotten truth that the utilization of a substrate by a certain type of bacterium depends on the substrate concentration and other environmental parameters. In particular this has become evident for hydrogen sulphide (see Sections 5.2.2 and 5.4).

Up to about 1960, knowledge of the biochemical details of the metabolism of inorganic electron donors was virtually non-existent. Since then

Table 5.1 Photosynthetic utilization of inorganic electron donors by the various types of phototrophic bacteria (i S = intracellular sulphur; e S = extracellular sulphur).

	H_2O	H_2	H_2S	$S_2O_3^{2-}$
Cyanobacteria	+	Several strains hydrogenase-positive; CO_2-photoreduction demonstrated in a few cases	Very much strain-dependent (see 5.4)	Only a few strains studied (see Section 5.4)
Rhodospirillaceae	–	Many strains grow with H_2 as electron donor (see 5.2.3)	Very much strain-dependent (see 5.2.2); some species: $S^{2-} \to$ e S; other species $S^{2-} \to SO_4^{2-}$	*Rhodopseudomonas palustris* and *Rp. sulfidophila* (no sulphur accumulation)
Chromatiaceae	–	Many strains hydrogenase-positive; some strains shown to grow with H_2 as electron donor	All strains: $S^{2-} \to$ i S $\to SO_4^{2-}$ except *Ectothiorhodospira* species ($S^{2-} \to$ e S $\to SO_4^{2-}$)	Several species (sulphur accumulation)
Chlorobiaceae	–	Several strains hydrogenase-positive; some shown to grow with H_2 as electron donor	All strains: $S^{2-} \to$ e S $\to SO_4^{2-}$	Some strains
Chloroflexaceae	–	Growth of *Chloroflexus aurantiacus* demonstrated (R. Sirevåg, unpublished)	One strain studied: $S^{2-} \to$ e S $\to$?	No growth

considerable progress has been made, but our knowledge still leads to a fragmentary picture. Important questions remain to be answered before certain, very basic, phenomena can be explained.

5.2 INORGANIC AND C_1-COMPOUNDS AS PHOTOSYNTHETIC ELECTRON DONORS

5.2.1 What substrates are utilized as electron donors and how?

Anaerobic degradation of organic compounds in nature leads to the formation of various oligo-carbon compounds, methane, sulphide, ammonia, hydrogen, and CO_2 as major products. Any of the reduced products could, at least theoretically, be used as an electron donor for photosynthesis. Two factors may be expected to be of extreme importance for the question whether utilization is likely to occur in nature, namely the redox potential of the substrate and the feasibility of anaerobic oxidation of the substrate.

Photosynthetic electron donors serve mainly to reduce CO_2 to cell material. For this purpose $NAD(P)^+$ and/or ferredoxin have to be reduced. This might be effected in three ways: directly; *via* the photochemical electron transport system by means of non-cyclic electron flow; or *via* reversed electron transport driven by the proton motive force (see Chapter 4). In this respect, the redox potential of the electron donor compared with that of NAD^+ or ferredoxin may be expected to be important. A coupling with an electron transport chain in any case excludes compounds with E'_0 values of about $+600\,mV$ or higher as substrates since the highest redox potentials of the components of the chains in obligately anoxygenic phototrophs are about $+450\,mV$ (see Chapters 3 and 4). See Table 5.2 for standard redox potentials of some important redox couples.

The actual redox potential of the electron donor may well be considerably different from the standard value, because concentrations of solutes in nature are usually much lower than the standard concentrations and the pH may differ from 7. To give one example: the standard value of H^+/H_2 is $-414\,mV$. This value is found at a hydrogen partial pressure of 1 atm (101325 Pa) which means a hydrogen concentration of 0.9 mM at 20°C. Actual hydrogen concentrations in anaerobic natural waters and sediments were found to range from not detectable in most cases to 3×10^{-3} mM (Winfrey *et al.*, 1977). At a concentration of 10^{-4} mM the actual redox potential is about $-300\,mV$. Such a value is considerably less favourable than the standard value, but would still be sufficiently low for a direct reduction of NAD^+.

Phototrophs that oxidize hydrogen sulphide to elemental sulphur as an intermediate or end product are not likely to encounter total sulphide concentrations of more than a few mM in nature; the redox potential of the S/HS^- couple at pH 7 will be about $-180\,mV$ rather than $-270\,mV$. Such a

Table 5.2 Redox potentials of some important couples. Data from Thauer et al. (1977) or calculated from the thermodynamic data in their paper.

	E'_0 (mV)
SO_4^{2-}/HSO_3^-	-516
CO_2/formate	-432
H^+/H_2	-414
$S_2O_3^{2-}/HS^- + HSO_3^-$	-402
$NAD^+/NADH$	-320
HS^-/S	-270
$HCHO/CH_3OH$	-180
HSO_3^-/HS^-	-116
adenosine-phosphosulphate/$AMP + HSO_3^-$	-60
$S_4O_6^{2-}/S_2O_3^{2-}$	$+24$
fumarate/succinate	$+33$
CH_3OH/CH_4	$+169$

value practically excludes a direct reduction of NAD^+ or oxidized ferredoxin by sulphide.

A suitable redox potential does not necessarily imply that phototrophs have learnt to deal with a substrate as an electron donor. Since CH_4 is an abundant product of anaerobic degradation, one might expect the existence of phototrophs that can utilize methane as an electron donor. Wertlieb and Vishniac (1967) reported the incorporation of $^{14}CH_4$ into cell material by a strain of *Rhodopseudomonas gelatinosa* but in view of unsuccessful attempts to obtain good enrichments of methane-utilizing phototrophic bacteria (Quayle & Pfennig, 1975; Hansen, unpublished) it seems unlikely that methane can be used by phototrophs. One may speculate why this is so. In the aerobic methane-oxidizers the first attack on the methane molecule is catalysed by a mixed-function oxygenase (Colby et al., 1979). Denitrifying methane-oxidizing bacteria have not been isolated. An efficient methane dehydrogenase which would be required in anaerobes may well be beyond the biochemical feasibilities despite the reasonable E'_0 of CH_3OH/CH_4 ($+169$ mV). Methanol dehydrogenation is not difficult and methanol utilization by phototrophs is possible, though most likely of little significance ecologically (see Section 5.2.4).

Bacteria that can utilize ammonia as a photosynthetic electron donor and form dinitrogen are a missing link in nature (Broda, 1977). The E'_0 of N_2/NH_4^+ is about -280 mV. The reverse reaction, catalysed by nitrogenase, functions with reducing power of about -0.4 V which results in a negative ΔG. Yet, the process requires a large amount of ATP (see Chapter 7). This indicates how difficult the reaction is. It is therefore not surprising that photolithotrophy based upon ammonia oxidation to nitrogen has not been discovered. Anaerobic oxidation to NO_2^- via NH_2OH is impossible: the E'_0 of

NH_2OH/NH_3 is too high ($+899\,mV$) to allow dehydrogenation of ammonia with an electron transport chain component as acceptor.

CO is produced not only by human activities but also in natural processes (Uffen, 1981). The E'_0 of CO_2/CO is very favourable ($-540\,mV$). Utilization of CO by phototrophic bacteria has been demonstrated unequivocally but not as a photosynthetic electron donor. Uffen (1976) discovered another mode of growth on CO, namely anaerobically in the dark. This was found with two strains of *Rp. gelatinosa* (Dashekvicz & Uffen, 1979). Growth was CO-dependent and based on the reaction:

$$CO + H_2O \rightarrow CO_2 + H_2 \qquad \Delta G'_0 = -20.1\,kJ$$

The only conceivable way in which biologically useful energy can be derived from this process would be the generation of a proton motive force as the result of membrane-associated CO dehydrogenation. Utilization of CO as a photosynthetic electron donor cannot be excluded. Hirsch (1968) isolated purple bacteria that grew under a CO atmosphere in the absence of significant amounts of organic substrates. Unfortunately, no follow-up studies on phototrophic growth on CO have appeared.

5.2.2 Reduced sulphur compounds as electron donors

Diversity of sulphide metabolism and organisms involved

During the past 15 years it has become clear that a far greater diversity of phototrophic bacteria can utilize sulphide as an electron donor than previously thought. Four species of the genus *Ectothiorhodospira* were described; these purple bacteria (assigned to the Chromatiaceae) oxidize sulphide to sulphate with extracellular sulphur as an intermediate, unlike, e.g. *Chromatium* which forms intracellular sulphur.

A sharp division between the Chromatiaceae and the purple nonsulphur bacteria (the Rhodospirillaceae) on the basis of sulphide utilization has become very difficult. *Rp. sulfidophila* is a marine purple bacterium that oxidizes sulphide to sulphate without detectable sulphur formation (Hansen & Veldkamp, 1973). It tolerates about the same sulphide concentrations as does *Chr. vinosum* but was assigned to the genus *Rhodopseudomonas* because of its similarity to *Rp. capsulata* in other respects.

Sulphide is a very toxic substrate; even most of the Chromatiaceae tolerate only 2–4 mM, some strains even less (Pfennig, 1977). Utilization of sulphide by common Rhodospirillaceae was discovered when sulphide was added in very low concentrations. Growth of *Rp. capsulata* Kb 1 (DSM 155) with sulphide is rapid (see Table 5.3) when sulphide concentrations are very low but completely inhibited at about 2 mM sulphide (Hansen & van Gemerden, 1972).

Rp. palustris can also utilize sulphide as an electron donor but is completely inhibited by even lower concentrations (less than 0.5 mM). The method of choice for obtaining dense cultures is the use of sulphide-limited chemostats. As during growth on thiosulphate, no elemental sulphur is formed, sulphate being the only product. In contrast, extracellular elemental sulphur was the only product detected in sulphide-grown *Rp. capsulata* cultures even at low dilution rates in the chemostat. *Rhodomicrobium vannielii* tolerates moderate sulphide concentrations (3 mM); in sulphide-limited chemostat cultures tetrathionate is the main product (Hansen, 1974). Sulphide utilization by this species had been reported earlier (van Niel, 1963) but was not generally accepted. The newly isolated purple bacterium strain 51 is very similar to a Rhodospirillacea such as *Rp. capsulata* (short rods to cocci in chains, phototrophic growth with several organic substrates, vitamin requirements, good aerobic growth); the products formed during growth with sulphide, however, are extracellular elemental sulphur and sulphate (Hansen *et al.*, 1975), the same as those of *Ectothiorhodospira*.

The green sulphur bacteria such as *Chlorobium* form extracellular sulphur and sulphate (cf. below). The gliding thermophilic green bacterium *Chloroflexus aurantiacus* grows best as a photoheterotroph. Photoautotrophic growth with sulphide is much slower (Madigan & Brock, 1975). About 70% of the sulphide oxidized was accounted for as elemental sulphur; virtually no

Table 5.3 Maximum specific growth rates observed under photoautotrophic conditions.
[1] sulphide + thiosulphate in case of *Chlorobium* 6430.
References: a. Hansen (1974); b. Klemme (1968); c. Madigan & Gest (1979); d. Douthit & Pfennig (1976); e. Slater & Morris (1973); f. van Gemerden & Jannasch (1971); g. van Gemerden & Beeftink (1978); h. van Gemerden (1974); i. Lippert & Pfennig (1969).
Some of the values were estimated from published growth curves.
Rp. = *Rhodopseudomonas*; *Rm.* = *Rhodomicrobium*; *Rs.* = *Rhodospirillum*; *Chr.* = *Chromatium*; *Chl.* = *Chlorobium*.

Strain	μ_{max} (h^{-1})			
	Sulphide[1]	Hydrogen	Formate	Methanol
Rp. sulfidophila W4	0.10[a]	0.07[a]	0.07[a]	
Rp. palustris DSM124	0.03[a]			
Rp. capsulata Kb1	0.14[a]	0.03[b]		
Rp. capsulata B10		0.20[c]		
Rp. acidophila 10050		0.17[d]		0.08[d]
Undesignated strain 51	0.20[a]			
Rm. vannielii ATCC17100	0.04[a]			
Rs. rubrum S1		0.017[e]		
Chr. vinosum DSM192	0.12[f]			
Chr. vinosum DSM185	0.12[g]			
Chr. weissei DSM171	0.04[h]			
Chl. limicola f. thiosulph. 6430	0.07[i]	0.03[i]		

sulphate was formed indicating that other products such as thiosulphate or tetrathionate may also be formed.

In Table 5.3 some data on specific growth rates are summarized.

Sulphide and thiosulphate utilization in batch and continuous cultures

In cultures of strains that form intracellular sulphur and sulphate, sulphur and sulphate formation are not strictly consecutive processes. In batch cultures of *Chr. vinosum* DSM 185, sulphate production begins while sulphide is being oxidized. The specific growth rate does not change when the sulphide is depleted (van Gemerden, 1968a; van Gemerden & Beeftink, 1978). In the first 'phase' of growth (that is, as long as sulphide is present), the flow of electrons from sulphide is faster than the flow from sulphur in the second phase. Expressed in equivalent units (see van Gemerden & Beeftink, 1978) the utilization rate of the electron donor during the first phase exceeds the rate of protein synthesis. The difference is accounted for by the accumulation of intracellular polyglucose. During the second phase, the rate of protein synthesis exceeds the rate of sulphur oxidation as a result of polyglucose breakdown (Fig. 5.1). In sulphide-limited continuous cultures of *Chromatium* species at low dilution rates intracellular sulphur concentrations are very low and most of the sulphide is oxidized to sulphate. The sulphur concentration increases at higher dilution rates (van Gemerden & Jannasch, 1971; van Gemerden, 1974). The potential rate of sulphide oxidation of continuous cultures, i.e. the rate found when the sulphide-limitation is relieved, is independent of the dilution rate. Acetate-limited continuous cultures had the same capacity for sulphide oxidation which indicates that the oxidation of sulphide to elemental sulphur is catalysed by a constitutive system (Beeftink & van Gemerden, 1979).

Information on the kinetics of sulphide oxidation by cultures of phototrophic bacteria that produce extracellular sulphur and sulphate is less extensive. In cultures of such bacteria sulphate formation does not start before sulphide has been used up. This was described for purple bacterium strain 51 (Hansen *et al.*, 1975) and for *Ectothiorhodospira mobilis* (Trüper, 1978). Hansen *et al.* (1975) suggested that sulphide and sulphur may compete for a common electron acceptor. Rapid reduction of the acceptor by high sulphide concentrations would prevent the oxidation of sulphur. At low dilution rates in chemostat cultures sulphide concentrations are very low and sulphate is the predominant product. *Chl. limicola f. thiosulfatophilum* excretes both elemental sulphur and thiosulphate during sulphide oxidation (Schedel, 1978). Sulphate production is not significant before sulphide has been completely consumed. Thiosulphate does not accumulate in cultures of the non-thiosulphate-utilizing *Chl. limicola* (Steinmetz & Fischer, 1981).

There are marked differences in the utilization of thiosulphate by phototrophs (for a review: Trüper, 1978). Purple sulphur bacteria such as *Chr. vinosum* and

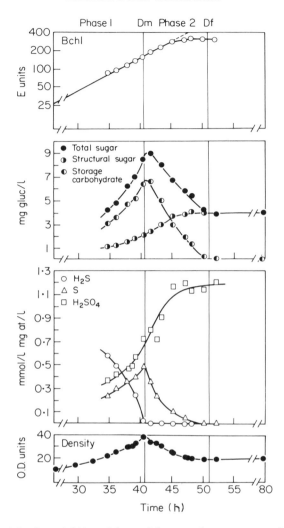

Fig. 5.1 Optical density, sulphide, sulphur, sulphate, total sugar, structural sugar, storage carbohydrate, and bacteriochlorophyll in a growing culture of *Chromatium*. Medium prepared with 1.20 mmol/l sulphide. All scales are linear, except that for chlorophyll, which is logarithmic. (Reproduced with permission from van Gemerden, 1968a)

Thiocapsa roseopersicina transiently accumulate intracellular sulphur. The sulphur is derived exclusively from the sulphane-S of thiosulphate. At low pH (6.25) thiosulphate is oxidized to tetrathionate which is not further metabolized. In cultures of *Rp. palustris* with high initial thiosulphate concentrations (10–20 mM) tetrathionate accumulates transiently (Rodova & Pedan, 1980). Oxidation of thiosulphate to sulphate by *Chl. limicola f. thiosulfatophilum* proceeds without the accumulation of sulphur or polythionates (Schedel, 1978) but *Chl. vibrioforme f. thiosulfatophilum* does form elemental sulphur (Trüper, 1978).

The conversion of sulphide to elemental sulphur and to sulphate

Virtually all sulphide-utilizing phototrophs can convert sulphide to detectable elemental sulphur. Because of the E'_0 of S/HS^- and the low concentrations of sulphide in the natural environment reduction of $NAD(P)^+$ or ferredoxin by sulphide is energy-requiring (see Section 5.2.1). It is not surprising, therefore, that all cell-free systems used so far in the study of sulphide oxidation to elemental sulphur contained membrane fractions or components that are thought to occur membrane-associated *in vivo*.

Identification of the *in vivo* acceptor of the electrons from sulphide is difficult. Sulphide is a reactive substrate and reduction of an isolated component by sulphide does not necessarily mean that the component is the only or major acceptor of the electrons in the cell. Cytochromes seem to play a key role in the oxidation of sulphide. For a review on cytochromes from phototrophic bacteria see Bartsch (1978). Table 5.4 summarizes data on cytochromes that may act as oxidants of sulphide.

One can only speculate about the biochemical differences that underlie the formation of intracellular and extracellular sulphur. The type of internal membrane system does not play a key role as evidenced by the different systems in *Rp. capsulata* and *Ectothiorhodospira* (both extracellular sulphur) and the similar systems in *Rp. capsulata* and *Chromatium*. The simplest explanation for the deposition of extracellular sulphur may be that the sulphide-oxidizing component (most probably a cytochrome) is located at the outer side of the cytoplasmic membrane and/or its invaginations. The photochemical apparatus and the electron transport chain are arranged in

Table 5.4 Components possibly involved in the oxidation of sulphide to sulphur.

	Organism	Product	Acceptor	References
Cytochrome c_{550}	*Thiocapsa roseopersicina*	S	?	Fischer & Trüper (1977)
Cytochrome c'	*Thiocapsa roseopersicina*	?	?	Fischer & Trüper (1979)
Flavocytochrome c_{552}	*Chr. vinosum*	?	?	Fukumori & Yamanaka (1979a)
Flavocytochrome c_{553}	*Chl. limicola* f. *thiosulfatophilum*	S(?)[a]	cyt c_{555}	Kusai & Yamanaka (1973a; b)
Flavocytochrome c_{553}	*Chl. limicola*	?	?	Steinmetz & Fischer (1981)

[a] Possibly $S_2O_3^{2-}$, see Trüper (1978)

such a way in the cytoplasmic membrane that the high potential components are at the outer side (see Chapter 4). This is true for Cyt c_2 which may be involved in sulphide oxidation in *Rp. capsulata* (Hansen & van den Boogaart, unpublished). One would expect a similar location of the flavocytochrome c_{553} in *Chlorobium*.

Are the intracellular sulphur globules of *Chromatium* formed intracytoplasmatically? In the aerobic sulphide-oxidizing bacterium *Beggiatoa* the intracellular sulphur is, in fact, located outside the cytoplasmic membrane in invaginations of it and surrounded by a special membrane with three electron-dense layers (Strohl & Larkin, 1978; Strohl *et al.*, 1982). Sulphur droplets isolated from *Chromatium* are surrounded by a thin (2.5 nm) membrane consisting of one type of protein with a molecular weight of 13 500 (Nicholson & Schmidt, 1971; Schmidt *et al.*, 1971). From published electron micrographs of thin sections it cannot be concluded where the sulphur globule is formed.

Intracellular sulphur can be reduced to sulphide in the dark in *Chromatium*. The reductant is NADH formed in the conversion of polyglucose to poly-β-hydroxybutyrate (van Gemerden, 1968b). The redox potential difference between $NAD^+/NADH$ and S/HS^- *in vivo* may be about 150 mV which should be sufficient for some energy conservation besides the ATP formed in glycolysis. It is not known whether the reduction of sulphur by NADH proceeds *via* exactly the reverse route of sulphide oxidation. *In vitro* flavocytochrome c_{552} can catalyse a non-stoichiometric reduction of elemental sulphur with reduced benzyl viologen (Fukumori & Yamanaka, 1979a).

The conversion of sulphide and elemental sulphur to sulphite is the least understood step in the formation of sulphate. Most probably the oxidation of sulphide to sulphite in purple sulphur bacteria is catalysed by a sulphite reductase that contains sirohaem. Such an enzyme has been isolated from *Chromatium vinosum* (Schedel *et al.*, 1979). It appeared to be rather similar to the bisulphite reductase from *Desulfovibrio*. The involvement of this enzyme in the oxidation of sulphide is suggested by its presence in sulphide-grown and absence from malate-grown cells. In the sulphite reductase assay with reduced methylviologen as electron donor considerable amounts of thiosulphate and trithionate were formed in addition to sulphide. It is not clear whether the *Chromatium* sulphite reductase can oxidize elemental sulphur. If not, the presence of a special sulphur-oxidizing enzyme may have to be postulated. Alternatively, sulphur may first have to be reduced to sulphide. Schedel (1978) did not detect a sulphite reductase in *Chl. limicola*.

The oxidation of sulphite to sulphate is relatively well understood in the phototrophic sulphur bacteria. As in some of the thiobacilli, sulphite is first oxidized to APS (adenosine-5′-phosphosulphate) in an AMP-dependent reaction catalysed by APS reductase:

$$AMP + SO_3^{2-} \rightarrow APS + 2e^- \qquad E_0' = -60\,mV$$

APS reductase was detected in various Chromatiaceae and Chlorobiaceae

(Trüper & Peck, 1970) but not in two Rhodospirillaceae that can oxidize sulphide or thiosulphate to sulphate, *Rp. palustris* (Trüper & Peck, 1970; Rodova & Pedan, 1980) and *Rp. sulfidophila* (Hansen & Veldkamp, 1973). Only low activities were found in *Ectothiorhodospira mobilis* (Trüper & Peck, 1970) and Kondratieva & Krasilnikova (1979) did not detect APS reductase in *E. shaposhnikovii* under any of the growth conditions tested. In the purple sulphur bacteria APS reductase is membrane-bound (Trüper & Peck, 1970). In *Chr. vinosum* the enzyme is tightly bound (Schwenn & Biere, 1979), while the APS-reductase of *Thiocapsa roseopersicina* readily leaches from the chromatophores and could be purified (Trüper & Rogers, 1971). It appeared to be unique in that it contained both a flavin and two haemes (a flavocytochrome). The APS reductases from sulphate-reducing bacteria, thiobacilli and, as was later shown, *Chlorobium* (Kirchhof & Trüper, 1974) are flavoproteins. The natural acceptor of the electrons in the APS reductase-catalysed oxidation of sulphite has not been identified in any of the phototrophic bacteria.

APS is converted to sulphate and ADP by ADP sulphurylase (Thiele, 1968). The AMP-dependent pathway of sulphate formation yields one ATP per sulphite oxidized to sulphate and not 0.5 ATP as suggested by Trüper (1978) and Ivanovskii & Petushkova (1977) (Table 5.5).

Table 5.5

$SO_3^{2-} + AMP$	→	$APS + 2e^-$	APS reductase
$APS + Pi$	→	$ADP + SO_4^{2-}$	ADP sulphurylase
$2ADP$	→	$AMP + ATP$	adenylate kinase
$SO_3^{2-} + ADP + Pi$	→	$ATP + SO_4^{2-} + 2e^-$	

Thiosulphate metabolism

Purple sulphur bacteria such as small *Chromatium* spp. and *Thiocapsa* utilize thiosulphate without a lag after growth with sulphide.

Thiosulphate oxidation is inducible in *Rp. sulfidophila* (Hansen, 1976) and *Rp. palustris* strains ATCC 11168 (Rolls & Lindstrom, 1967a) and Nakamura (Rodova & Pedan, 1980) after heterotrophic growth.

Three enzymes have been claimed to be involved in the initial step of thiosulphate metabolism, namely thiosulphate reductase:

$$S_2O_3^{2-} + 2H \rightarrow S^{2-} + SO_3^{2-} + 2H^+ \qquad \text{Eqn. 5.1}$$

rhodanese (thiosulphate: cyanide sulphur transferase E.C. 2.8.1.1):

$$S_2O_3^{2-} + CN^- \rightarrow SCN^- + SO_3^{2-} \qquad \text{Eqn. 5.2}$$

and thiosulphate: acceptor oxidoreductase:

$$2S_2O_3^{2-} + 2X \rightarrow 2X^- + S_4O_6^{2-} \qquad \text{Eqn. 5.3}$$

In *Rp. palustris* strain Nakamura all three enzymes were detected (Rodova & Pedan, 1980). Thiosulphate reductase (with dithiothreitol as H-donor) had by far the highest activity. Cell-free extracts of *Rp. palustris* ATCC 11168 catalyse Eqn. 5.3 with ferricyanide as acceptor only if the cells have been grown in the presence of thiosulphate (Rolls & Lindstrom, 1967b). A *c*-type cytochrome with $E'_0 = +228\,\text{mV}$ was reported to function as acceptor for the thiosulphate-oxidizing enzyme of *Rp. palustris* (Appelt et al., 1979).

Petushkova and Ivanovskii (1977) detected both thiosulphate reductase and rhodanese in extracts of *Thiocapsa roseopersicina* but negligible activities of thiosulphate:acceptor oxidoreductase. At pH 6.25 *Chr. vinosum* converts thiosulphate to tetrathionate which is not further metabolized (Smith, 1966). An enzyme which catalyses this oxidation was partially purified. Reduced HiPiP (high potential non-haem iron protein) is probably the natural electron acceptor (Fukumori & Yamanaka, 1979b). Schmitt et al. (1981), however, isolated a thiosulphate:acceptor oxidoreductase which functions with flavocytochrome c_{552} as acceptor.

It is not clear what product is formed in the oxidation of thiosulphate catalysed by the thiosulphate:cytochrome c_{551} oxidoreductase from *Chl. limicola* f. *thiosulfatophilum*. Polythionates were not formed (Kusai & Yamanaka, 1973b). The acceptor cytochrome c_{551} was not detected in the non-thiosulphate-utilizing *Chl. limicola* (Steinmetz & Fischer, 1981).

5.2.3 Hydrogen as an electron donor

Utilization of hydrogen by purple bacteria was discovered by Roelofsen (1935). Only during the last decades, however, has a wide range of species been studied. Klemme (1968) found that in enrichments with H_2/CO_2 *Rp. capsulata* strains grew best. Most strains of this species can grow with hydrogen as electron donor (Weaver et al., 1975). Good photoautotrophic growth with hydrogen was also reported for other species (see Table 5.3).

Most *Chlorobium* strains are hydrogenase-positive. The green sulphur bacteria cannot reduce sulphate and therefore, during growth with hydrogen a reduced sulphur source is required. To some extent cysteine can be used as such but for prolonged good growth sulphide is needed (Lippert, 1967; Lippert & Pfennig, 1969).

Hydrogenases

Extremely low concentrations of hydrogen (below $10^{-4}\,\text{mM}$) would make a direct reduction of NAD^+ by hydrogen thermodynamically unfavourable (and even more so at high $NADH/NAD^+$ ratios) (see Section 5.2.1). In none of the phototrophic bacteria studied has an enzyme been demonstrated that catalyses such a direct reduction. NAD^+-dependent hydrogenases occur in a few knallgasbacteria (Schneider & Schlegel, 1977); in most of them only an NAD^+-independent membrane-bound enzyme is present.

Present knowledge indicates that the hydrogenases of the purple bacteria are membrane-bound or at least membrane-associated. In *Rp. capsulata* strains Kb1 (Klemme, 1969) and B10 (Colbeau et al., 1978) the hydrogenase is tightly membrane-bound but can be solubilized with detergents. It is cold-labile and not very sensitive to oxygen. Hydrogenase activity of *Rp. capsulata* cells is greatest in cells grown with hydrogen as the electron donor or photoheterotrophically with a 'poor' nitrogen source (glutamate) (Colbeau et al., 1980). The latter conditions lead to the induction of nitrogenase and hydrogen evolution. The observations therefore suggest that hydrogenase is inducible.

Adams and Hall (1978; 1979) described the solubilization and purification of a membrane-bound hydrogenase from *Rs. rubrum*. According to Hiura et al. (1979) the greater part of the hydrogenase from this species is released into the culture medium during growth. The purified enzyme had a molecular weight of about 65 000 and most probably consisted of a single polypeptide chain with four iron atoms and four acid-labile sulphur atoms per molecule (Adams & Hall, 1978). It did not use NAD^+ or $NADP^+$ as electron acceptors and did not readily interact with *Rs. rubrum* ferredoxins. Gogotov and Laurinavichene (1976) reported reduction of NAD^+ by hydrogen in extracts of *Rs. rubrum*. This reduction was not energy-dependent in contrast to that observed by Klemme (1967) in chromatophores from *Rp. capsulata*. The effect of ferredoxin on NAD^+-reduction was studied as follows: a membrane-bound ferredoxin was solubilized, purified and added to the assay system with cell-free extract from which soluble ferredoxins had been removed. This gave a 50% stimulation of NAD^+ reduction.

Gogotov et al. (1978a; 1978b) purified a hydrogenase from the soluble fraction of *Thiocapsa roseopersicina* and one which was solubilized from the chromatophore fraction. The enzymes had the same molecular weights (68 000), similar contents of iron and acid-labile sulphur (about 3.9 atoms per molecule), similar redox potentials ($-280\,mV$) and similar amino acid compositions and may in fact be identical. The hydrogenase does not use $NAD(P)^+$ or FAD and FMN as acceptors and the authors suggested that c-type cytochromes performed this function. Contradictory results have been obtained with the hydrogenase(s) from *Chr. vinosum* (see Llama et al., 1981). According to Gitlitz and Krasna (1975) the purified membrane-bound hydrogenase, which contains 4 iron and 4 acid-labile sulphur atoms per molecule, does not reduce ferredoxin or NAD^+ with hydrogen. In a crude system Buchanan and Bachofen (1968) observed the ferredoxin-dependent reduction of NAD^+ (and at a lower rate, of $NADP^+$) with hydrogen. For optimal activity both the soluble and the chromatophore fractions were required.

There are no detailed reports on the hydrogenases of green sulphur bacteria.
Several aspects of the process of overall reduction of $NAD(P)^+$ or ferredoxin with hydrogen in purple bacteria are still poorly understood. In this respect the situation is not very different from 1962 when Ormerod and Gest (1962) stated: '... but much remains to be investigated particularly in connection with the basic problem of the fate of electrons derived from the oxidation of H_2'. The identity of the natural electron acceptor(s) and the localization in the membrane of the hydrogenases have not yet been studied in detail. The finding of Hiura et al., (1979) described above may suggest a localization at the periplasmic side of the cytoplasmic membrane and its invaginations.

5.2.4 Methanol and formate as electron donors

Utilization of methanol was reported for strains belonging to five species of Rhodospirillaceae. Reasonable growth rates (see Table 5.3), however, were only obtained with a few Rp. acidophila strains (Douthit & Pfennig, 1976). Formate is a suitable electron donor for Rp. palustris (low growth rates), Rp. acidophila and Rp. sulfidophila.

Methanol and formate have been shown to be true electron donors. Bacterial assimilation of reduced C_1-compounds occurs mainly at the level of formaldehyde (Colby et al., 1979). On the other hand, assimilation of methanol or formate in purple bacteria is essentially autotrophic, that is at the level of CO_2. This was demonstrated by short-term labelling experiments and measurement of key enzymes of the Calvin cycle (Stokes & Hoare, 1969; Sahm et al., 1976). Methanol or formate are oxidized to CO_2 and the reducing power obtained is used for CO_2-fixation.

Rp. acidophila converts methanol to formaldehyde with an $NAD(P)^+$-independent alcohol dehydrogenase which can be linked to phenazine-methosulphate/2,6-dichlorophenolindophenol (Sahm et al., 1976). It is rather similar to the methanol dehydrogenases from aerobic methylotrophs (Bamforth & Quayle, 1978) and has the same pyrrolo-quinoline quinone as its prosthetic group (Duine et al., 1980). There are, however, some marked differences: the alcohol dehydrogenase of Rp. acidophila is, as Bamforth and Quayle (1978) put it, a higher alcohol dehydrogenase by design and a methanol dehydrogenase by accident. It is normally involved in the oxidation of, for instance, ethanol (Krasilnikova, 1976; Bamforth & Quayle, 1979) with which it shows a higher V_{max} than with methanol; it also has a lower K_m for ethanol (0.006–0.03 mM) than for methanol (12 mM). The enzyme is probably loosely attached to the membranes. Presumably the proton-motive force drives the electron flow to NAD^+ from the electron acceptor in the membrane which has not yet been identified. Photoautotrophic growth of Rp. acidophila with hydrogen is much faster than growth on methanol. This suggests that the oxidation of methanol to CO_2 is rate-limiting (Douthit & Pfennig, 1976).

Oxidation of formaldehyde to formate in *Rp. acidophila* may be effected by the methanol dehydrogenase or by an NAD^+- and glutathione-dependent formaldehyde dehydrogenase (Sahm *et al.*, 1976). Growth on methanol induces the formaldehyde dehydrogenase.

Formate is oxidized to CO_2 in an NAD^+-dependent reaction in *Rp. acidophila* grown on methanol or formate (Sahm *et al.*, 1976) and in *Rp. palustris* grown on formate (Yoch & Lindstrom, 1969).

5.3 CHEMOLITHOTROPHY

5.3.1 Reduced sulphur compounds as energy source

The first example of facultative chemolithotrophy among the phototrophic bacteria was the discovery that *Thiocapsa roseopersicina* BBS can grow aerobically in the dark on CO_2 with thiosulphate as energy source (Bogorov, 1974). Growth under these conditions is slower than in the light (Kondratieva *et al.*, 1976). Cells in well-aerated cultures contain only traces of bacteriochlorophyll and virtually no chromatophores. The levels of ribulose-bisphosphate carboxylase are almost the same as anaerobically in the light. Sulphur accumulates intracellularly during the oxidation of thiosulphate to sulphate.

Amoebobacter roseus can also grow chemoautotrophically but only at low oxygen tensions (Gorlenko, 1974). In a survey comprising 45 strains of 17 species Kämpf and Pfennig (1980) studied the capacity of Chromatiaceae for chemoautotrophic, mixotrophic and organotrophic growth in the dark, especially at low oxygen tensions. They demonstrated micro-aerobic chemoautotrophic growth with thiosulphate and sulphide of *Chr. vinosum*, *Chr. minus*, *Chr. violascens*, *Chr. gracile*, *Thiocystis violacea* and additional strains of *Thiocapsa roseopersicina* and *Amoebacter roseus*. The large *Chromatium* sp., *Thiospirillum* and four other species did not grow as micro-aerobic chemautotrophs. None of the green sulphur bacteria tested grew in the dark.

The rather oxygen-tolerant *Thiocystis violacea* 2311 formed 11.3 g cell material per mol thiosulphate consumed, indicating a surprisingly high efficiency of the lithotrophic energy generation. Kuenen (1979) reported values of about half this for thiosulphate-limited chemostat cultures of *Thiobacillus neapolitanus* and *Thiobacillus* A2. Yields of *Thiobacillus denitrificans* are about as high as those of *Thiocystis* (Timmer-ten Hoor, 1976). Part of the explanation for the different efficiencies may be that both *Thiobacillus denitrificans* and *Thiocystis violacea* possess APS-reductase for the oxidation of sulphite whereas this enzyme is absent in the other organisms. Oxidation of sulphite to APS allows additional substrate-level phosphorylation (see Section 5.2.2).

Clearly, the (micro-) aerobic oxidation of the reduced sulphur compounds is not a marginal process in these phototrophs in terms of energy yield.

Maximal respiration rates of *Thiocystis violacea* with thiosulphate are so low (about 70 μl O_2 mg^{-1} protein h^{-1}) that only very low growth rates can be expected. The specific growth rate was 0.023 h^{-1} which is only about 5% of the μ_{max} of obligately chemolithotrophic thiobacilli.

Two species of Rhodospirillaceae oxidize thiosulphate aerobically, namely *Rp. palustris* (Rolls & Lindstrom, 1967a; Rodova & Pedan, 1980) and *Rp. sulfidophila* (Hansen, 1974). Thiosulphate is oxidized at a reasonable rate only by cells that have been grown in the presence of thiosulphate (or sulphide) either phototrophically or chemotrophically. Incorporation of thiosulphate into pyruvate medium did not increase the yield in aerobic batch cultures of either species. In L-malate-limited chemostat cultures of *Rp. sulfidophila*, however, a significant increase in cell density was observed when thiosulphate was supplied (Hansen, 1979; unpublished). Aerobic thiosulphate oxidation generates sufficient energy for survival. Thiosulphate addition to aerobically incubated cell suspensions prevented a rapid decline in viability.

5.3.2 Hydrogen, formate and methanol as sources of energy

Since there is no basic difference in the utilization of hydrogen or formate and methanol as energy sources, these organic substrates are included in this discussion of chemolithotrophy.

Cell suspensions of *Rp. capsulata* can carry out an oxygen- and hydrogen-dependent fixation of $^{14}CO_2$ in the dark (Stoppani et al., 1955). Aerobic autotrophic *growth* of a purple bacterium with hydrogen was, however, first demonstrated by Pfennig and Siefert (1977) with *Rp. acidophila* and later described for *Rp. capsulata* strains B10 (Madigan & Gest, 1979) and Kb1 (Siefert & Pfennig, 1979) and *Rp. sulfidophila* (Hansen, 1979). In all cases reduced oxygen tensions were required. Doubling times of 6 and 9 h were reported for *Rp. capsulata* strains B10 and Kb1 respectively. It is intriguing that the anaerobic phototrophic growth of the latter strain on H_2/CO_2 is much slower ($\mu = 0.03$ h^{-1}; see Table 5.3). Chemoautotrophic growth of strain Kb1 is possible at pO_2 values up to 13 kPa. The inhibitory effect of oxygen was ascribed to a suppression of hydrogenase. In *Rp. capsulata* the electrons from hydrogen can flow to oxygen *via* either of the terminal oxidases as shown by mutant studies (Madigan & Gest, 1979) and the effect of KCN on the hydrogen-dependent respiration of chromatophores (Paul et al., 1979). *Rp. acidophila* is more sensitive to oxygen during chemotrophic growth with H_2 and CO_2; it did not grow at a pO_2 of 3.3 kPa. Growth on methanol was possible at $pO_2 = 20$ kPa, but on formate only at very low pO_2. A biochemical explanation for this difference is not obvious since methanol is oxidized to CO_2 *via* formate. Chemotrophic growth on methanol is slow ($\mu = 0.02$ h^{-1}).

5.4 ANOXYGENIC PHOTOSYNTHESIS OF CYANOBACTERIA

It has long been known that many cyanobacteria can thrive anaerobically or at very low oxygen concentrations under reducing conditions. It was not until 1975, however, that anoxygenic photosynthesis of a cyanobacterium with H_2S as the electron donor was demonstrated in experiments with a strain of *Oscillatoria limnetica* isolated from a H_2S-rich layer of the Solar Lake near Elat (Israel; for reviews: Padan, 1979a; b). Low concentrations of sulphide (0.2 mM) strongly (95%) inhibit CO_2 fixation of *O. limnetica*. In the presence of higher concentrations of sulphide (3 mM), however, CO_2 fixation resumes at a considerable rate after an adaptation period of about two hours during which protein synthesis is required. This process is not inhibited by DCMU [3-(3,4-dichlorophenyl)-1,1,-dimethylurea] in contrast to oxygenic photosynthesis, indicating that photosystem (PS) II is not involved (Cohen et al., 1975a; Oren & Padan, 1978). Sulphide is oxidized to extracellular elemental sulphur (Cohen et al., 1975b):

$$2 H_2S + CO_2 \rightarrow (CH_2O) + 2 S + H_2O$$

A comparison of quantum yield spectra in the presence and absence of H_2S showed unequivocally that only PSI is involved in anoxygenic photosynthesis (Oren et al., 1977). Only anoxygenic photosynthesis can be driven effectively by 703 nm light.

Among the cyanobacteria photosynthesis with H_2S as electron donor is not restricted to *O. limnetica*. In a study of Garlick et al. (1977) 11 out of 20 cyanobacterial strains belonging to five genera were found to perform anoxygenic photosynthesis with 703 nm light in the presence of DCMU. The maximal CO_2 fixation rates observed varied considerably. Rates for *Anacystis nidulans* never exceeded about 1% of the rate of oxygenic CO_2 photoassimilation (Peschek, 1978). In the absence of DCMU the optimal sulphide concentration (0.1 mM) hardly affected oxygenic photosynthesis. The range of sulphide concentrations supporting CO_2 fixation depends upon the cyanobacterial strain (Fig. 5.2). The figure shows that the strain that grows at the highest sulphide concentrations (*O. limnetica*) has a poor affinity for sulphide. Obligately anoxygenic phototrophs such as *Chromatium* and *Rp. capsulata* have K_s values for sulphide in the micromolar range (see Chapter 8). Nothing is known about the biochemistry of sulphide oxidation in cyanobacteria.

Anaerobically in the dark *O. limnetica* uses sulphur as an electron acceptor in the oxidation of polyglucose to CO_2 (Oren & Shilo, 1979). In the presence of sulphur only a small amount of lactate is formed (Oren & Shilo, 1979). Lactate is normally a major product of anaerobic polyglucose degradation. Since the route for complete polyglucose degradation (the oxidative pentose phosphate cycle) does not yield ATP, most of the ATP formation in the presence of sulphur must occur in the reduction of sulphur with NADPH. With various

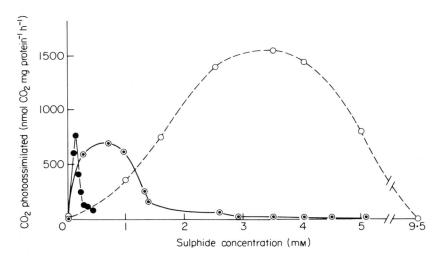

Fig. 5.2 Photoassimilation rate of Lyngbya (7104) and *A. halophytica* (7418) as a function of the Na_2S concentration. The rate of CO_2 photoassimilation at saturating intensity ($0.9\,\mu E\,cm^{-2}\,min^{-1}$) of 703 nm light, in the presence of DCMU ($5\,\mu M$), as measured at various Na_2S concentrations as described in Materials and Methods. Initial concentrations of Na_2S indicated in the figure were determined by the methylene blue photometric method. (●) *Lyngbya* (7104); (⊙) *Aphanothece halophytica* (7418); (○) *O. Limnetica*. (Reproduced with permission from Garlick *et al*, 1977).

assumptions it was suggested that about 3.5 ATP are formed per glucose unit.

Anoxygenic photosynthesis has also been observed with thiosulphate which is not normally an inhibitory substrate. Utkilen (1976) demonstrated thiosulphate-dependent anoxygenic photosynthesis by *Anacystis nidulans* 14011 in the presence of DCMU. In the absence of DCMU at low light intensities, thiosulphate reduced the doubling time of a culture to that obtained with saturating light intensities. At a light intensity of 20 klux thiosulphate had no effect on the CO_2 fixation rate of cells. At 1 klux with thiosulphate the rate was 39% of that at 20 klux whereas no measurable CO_2 fixation occurred in its absence. Thiosulphate was oxidized to sulphate. Obviously, the effect of the energetically more favourable use of thiosulphate as an electron donor becomes most pronounced at low light intensities. With a different *Anacystis* strain (1402–1) Peschek (1978) detected a very low rate of thiosulphate-dependent CO_2 fixation in the presence of DCMU; growth with thiosulphate in the presence of DCMU was not obtained. Clearly, more work needs to be done on thiosulphate utilization with more strains before the importance of this substrate can be properly evaluated.

When DCMU was used to block PS II, *Aphanothece* 7418 and *O. limnetica* were able to photoreduce CO_2 with hydrogen at rates that were about 40 and 6%, respectively, of those of sulphide-dependent photosynthesis (Belkin &

Padan, 1978). Hydrogen-dependent anaerobic phototrophic growth of *Anacystis* in the presence of DCMU was not possible; however, hydrogen induces the synthesis of a hydrogenase which can sustain a low rate of CO_2 fixation (Peschek, 1978; 1979a). *Anacystis* cannot fix dinitrogen; involvement of the hydrogenase in uptake of hydrogen lost through nitrogenase can therefore be excluded. *O. limnetica* can form nitrogenase under anaerobic conditions (Belkin *et al.*, 1979) so the main function of the hydrogenase of this organism may be recycling of hydrogen produced by the nitrogenase.

The utilization of electron donors other than water as discussed above underlines the special position of the cyanobacteria. Though usually oxygenic, some of them have maintained the option of anoxygenic photosynthesis which may confer a pronounced ecological advantage as illustrated in the case of *O. limnetica* (see Padan, 1979a).

5.5 CONCLUDING REMARKS

This chapter illustrates how fruitful a field of research the study of electron donor metabolism in phototrophic bacteria has been. In the past two decades several new types of anoxygenic phototrophic bacteria were discovered. Anoxygenic photosynthesis was recognized as an alternative mode of growth of some cyanobacteria. Chemolithotrophy emerged as a new aspect of the physiology of several purple bacteria. Considerable progress has been made in elucidating the biochemical details of the metabolism of electron donors. However, wide gaps remain, especially in our knowledge of the metabolism of reduced sulphur compounds. A biochemical explanation for the formation of either intracellular or extracellular sulphur is lacking. The oxidation and reduction of sulphur are poorly understood. The role of thiosulphate reductase in thiosulphate metabolism requires further study. The biochemistry of sulphide oxidation in cyanobacteria is an example of an untouched field of research.

The phototrophic bacteria do not grow as rapidly and are not always as easy to handle as *Escherichia coli*. This, however, should not discourage microbial physiologists from making more extensive use of mutants when studying electron donor metabolism of phototrophs. The role of certain proteins in the metabolism of sulphide or thiosulphate could be elucidated more conclusively by employing mutants impaired in the metabolism of the electron donor.

REFERENCES

ADAMS M. W. & HALL D. O. (1978) Physical and catalytic properties of the hydrogenase of *Rhodospirillum rubrum*. In *Hydrogenases, their Catalytic Activity, Structure and Function* (Ed. by H. G. Schlegel & K. Schneider) pp. 159–69. Goltze, Göttingen.

ADAMS M. W. & HALL D. O. (1979) Properties of the solubilized membrane-bound hydrogenase from the photosynthetic bacterium *Rhodospirillum rubrum*. *Arch. Biochem. Biophys.*, **195**, 288–99.

APPELT N., WEBER H., WIELUCH S. & KNOBLOCH K. (1979) Ueber das System Thiosulfat-Cytochrom c-Oxidoreductase aus *Rhodopseudomonas palustris. Ber. Dt. Bot. Ges.*, **92**, 365–78.
BAMFORTH C. W. & QUAYLE J. R. (1978) The dye-linked alcohol dehydrogenase of *Rhodopseudomonas acidophila*. Comparison with dye-linked methanol dehydrogenases. *Biochem. J.*, **169**, 677–86.
BAMFORTH C. W. & QUAYLE J. R. (1979) Structural aspects of the dye-linked alcohol dehydrogenase of *Rhodopseudomonas acidophila. Biochem. J.*, **181**, 517–24.
BARTSCH R. G. (1978) Cytochromes. In *The Photosynthetic Bacteria* (Ed. by R. K. Clayton & W. R. Sistrom), pp. 249–79. Plenum Press, New York.
BEEFTINK H. H. & GEMERDEN H. VAN (1979) Actual and potential rates of substrate oxidation and product formation in continuous cultures of *Chromatium vinosum. Arch. Microbiol.*, **121**, 161–7.
BELKIN S., COHEN Y. & PADAN E. (1979) Sulfide-dependent activities of the cyanobacterium *Oscillatoria limnetica*—further observations. Proceedings of the third International Symposium on Photosynthetic Prokaryotes. Oxford, B17.
BELKIN S. & PADAN E. (1978) Hydrogen metabolism in the facultative anoxygenic cyanobacteria (blue-green algae) *Oscillatoria limnetica* and *Aphanothece halophytica. Arch. Microbiol.*, **116**, 109–11.
BOGOROV L. V. (1974) About the properties of *Thiocapsa roseopersicina* strain BBS, isolated from the estuary of the White Sea. *Microbiology*, **43**, 275–80.
BRODA E. (1977) Two kinds of lithotrophs missing in nature. *Z. Allg. Mikrobiol.*, **17**, 491–3.
BUCHANAN B. B. & BACHOFEN R. (1968) Ferredoxin dependent reduction of NAD with hydrogen gas by subcellular preparations from the photosynthetic bacterium *Chromatium. Biochim. Biophys. Acta*, **162**, 607–10.
COHEN Y., PADAN E. & SHILO M. (1975a) Facultative anoxygenic photosynthesis in the cyanobacterium *Oscillatoria limnetica. J. Bact.*, **123**, 855–61.
COHEN Y., JØRGENSEN B. B., PADAN E. & SHILO M. (1975b) Sulfide-dependent anoxygenic photosynthesis in the cyanobacterium *Oscillatoria limnetica. Nature, Lond.*, **257**, 489–92.
COLBEAU A., CHABERT J. & VIGNAIS P. M. (1978) Hydrogenase activity in *Rhodopseudomonas capsulata*. Stability and stabilization of the solubilized enzyme. In *Hydrogenases: their Catalytic Activity, Structure and Function* (Ed. by H. G. Schlegel & K. Schneider), pp. 183–97. Goltze, Göttingen.
COLBEAU A., KELLEY B. C. & VIGNAIS P. M. (1980) Hydrogenase activity in *Rhodopseudomonas capsulata*: relationship with nitrogenase activity. *J. Bact.*, **144**, 141–8.
COLBY J., DALTON H. & WHITTENBURY R. (1979) Biological and biochemical aspects of microbial growth on C_1 compounds. *Ann. Rev. Microbiol.*, **33**, 481–518.
DASHEKVICZ M. P. & UFFEN R. L. (1979) Identification of a carbon monoxide-metabolizing bacterium as a strain of *Rhodopseudomonas gelatinosa* (Molisch) van Niel. *Int. J. syst. Bact.*, **79**, 145–8.
DOUTHIT H. A. & PFENNIG N. (1976) Isolation and growth rates of methanol utilizing Rhodospirillaceae. *Arch. Microbiol.*, **107**, 233–4.
DUINE J. A., FRANK J. & VERWIEL P. E. (1980) Structure and activity of the prosthetic group of methanol dehydrogenase. *Eur. J. Biochem.*, **108**, 187–92.
FISCHER U. & TRÜPER H. G. (1977) Cytochrome c-550 of *Thiocapsa roseopersicina*: properties and reduction by sulfide. *FEMS Microbiol. Lett.*, **1**, 87–90.
FISCHER U. & TRÜPER H. G. (1979) Some properties of cytochrome c′ and other hemoproteins of *Thiocapsa roseopersicina. Curr. Microbiol.*, **3**, 41–4.
FUKUMORI Y. & YAMANAKA T. (1979a) Flavocytochrome c of *Chromatium vinosum*. Some enzymatic properties and subunit structure. *J. Biochem. Tokyo*, **85**, 1405–14.
FUKUMORI Y. & YAMANAKA T. (1979b) A high-potential nonheme iron protein (HiPiP)-linked thiosulfate-oxidizing enzyme derived from *Chromatium vinosum. Curr. Microbiol.*, **3**, 117–20.

GARLICK S., OREN A. & PADAN E. (1977) Occurrence of facultative anoxygenic photosynthesis among filamentous and unicellular cyanobacteria. *J. Bact.*, **129**, 623–9.
GEMERDEN H. VAN (1968a) Utilization of reducing power in growing cultures of *Chromatium*. *Arch. Mikrobiol.*, **64**, 111–17.
GEMERDEN H. VAN (1968b) On the ATP generation by *Chromatium* in darkness. *Arch. Mikrobiol.*, **64**, 118–24.
GEMERDEN H. VAN (1974) Coexistence of organisms competing for the same substrate: an example among the purple sulfur bacteria. *Microb. Ecol.*, **1**, 104–19.
GEMERDEN H. VAN & BEEFTINK H. H. (1978) Specific rates of substrate oxidation and product formation in autotrophically growing *Chromatium* cultures. *Arch. Microbiol.*, **119**, 135–43.
GEMERDEN H. VAN & JANNASCH H. W. (1971) Continuous culture of Thiorhodaceae. Sulfide and sulfur-limited growth of *Chromatium vinosum*. *Arch. Mikrobiol.*, **79**, 345–53.
GITLITZ P. H. & KRASNA A. I. (1975) Structure and catalytic properties of hydrogenase from *Chromatium*. *Biochemistry, N.Y.*, **14**, 2561–7.
GOGOTOV I. N. & LAURINAVICHENE T. V. (1976) The role of ferredoxin in the hydrogen metabolism in *Rhodospirillum rubrum*. *Microbiology*, **44**, 517–22.
GOGOTOV I. N., ZORIN N. A., SEREBRIAKOVA L. T. & KONDRATIEVA E. N. (1978a) The properties of hydrogenase from *Thiocapsa roseopersicina*. *Biochim. Biophys. Acta*, **523**, 335–43.
GOGOTOV I. N., ZORIN N. A. & LAURINAVICHENE T. V. (1978b) Comparison of properties and function of hydrogenase and NADP-reductase isolated from *Thiocapsa roseopersicina*. In *Hydrogenases, their Catalytic Activity, Structure and Function* (Ed. by H. G. Schlegel & K. Schneider), pp. 171–82. Goltze, Göttingen.
GORLENKO V. M. (1974) The oxidation of thiosulfate of *Amoebobacter roseus* in the dark under microaerophilic conditions. *Microbiology*, **43**, 624–5.
HANSEŃ T. A. (1974) Sulfide als electrondonor voor Rhodospirillaceae. Thesis. University of Groningen.
HANSEN T. A. (1976) Some aspects of the oxidation of reduced sulfur compounds by *Rhodopseudomonas sulfidophila*. In Proceedings of the Second International Symposium on Photosynthetic Prokaryotes (Ed. by G. A. Codd & W. D. P. Stewart), p. 43. Dundee.
HANSEN T. A. (1979) Aerobic growth of *Rhodopseudomonas sulfidophila* in the dark and in the light. In Abstracts of the third International Symposium on Photosynthetic Prokaryotes (Ed. by J. M. Nichols), B26. Oxford.
HANSEN T. A. & VAN GEMERDEN H. (1972) Sulfide utilization by purple nonsulfur bacteria. *Arch. Mikrobiol.*, **86**, 49–56.
HANSEN T. A., SEPERS A. B. J. & VAN GEMERDEN H. (1975) A new purple bacterium that oxidizes sulfide to extracellular sulfur and sulfate. *Pl. Soil*, **43**, 17–27.
HANSEN T. A. & VELDKAMP H. (1973) *Rhodopseudomonas sulfidophila*, nov. spec., a new species of the purple nonsulfur bacteria. *Arch. Mikrobiol.*, **92**, 45–58.
HIRSCH P. (1968) Photosynthetic bacterium growing under carbon monoxide. *Nature, Lond.*, **217**, 555–6.
HIURA H., KAKUNO T., YAMASHITA J., BARTSCH R. G. & HORIO T. (1979) Extracellular hydrogenase from photosynthetic bacterium, *Rhodospirillum rubrum*. *J. Biochem., Tokyo*, **86**, 1151–3.
IVANOVSKII R. N. & PETUSHKOVA YU. P. (1977) Substrate phosphorylation during oxidation of sulfite by *Thiocapsa roseopersicina* depending on conditions of growth. *Microbiology*, **45**, 941–2.
KÄMPF C. & PFENNIG N. (1980) Capacity of Chromatiaceae for chemotrophic growth. Specific respiration rates of *Thiocystis violacea* and *Chromatium vinosum*. *Arch. Microbiol.*, **127**, 125–35.
KIRCHHOFF J. & TRÜPER H. G. (1974) Adenylylsulfate reductase of *Chlorobium limicola*. *Arch. Microbiol.*, **100**, 115–20.
KLEMME J. H. (1967) Hydrogenase und photosynthetischer Elektronentransport bei *Rhodospirillum rubrum* und *Rhodopseudomonas capsulata*. Thesis. University of Göttingen.

KLEMME J. H. (1968) Untersuchungen zur Photoautotrophie mit molekularem Wasserstoff bei neuisolierten schwefelfreien Purpurbakterien. *Arch. Mikrobiol.*, **64**, 29–42.
KLEMME J. H. (1969) Reaktionen der Hydrogenase aus *Rhodopseudomonas capsulata* im partikelgebundenen und gelösten Zustand. *Z. Naturf.*, **24b**, 603–12.
KONDRATIEVA E. N. & KRASILNIKOVA E. N. (1979) Growth of *Ectothiorhodospira shaposhnikovii* on media with various sulfur compounds. *Microbiology*, **48**, 152–8.
KONDRATIEVA E. N., ZHUKOV V. G., IVANOVSKY R. N., PETUSHKOVA YU. & MONOSOV E. Z. (1976) The capacity of phototrophic sulfur bacterium *Thiocapsa roseopersicina* for chemosynthesis. *Arch. Microbiol.*, **108**, 287–92.
KRASILNIKOVA E. N. (1976) Alcohol dehydrogenase activity of nonsulfur purple bacteria. *Microbiology*, **44**, 716–20.
KUENEN J. G. (1979) Growth yields and 'maintenance energy requirement' in *Thiobacillus* species under energy limitation. *Arch. Microbiol.*, **122**, 183–8.
KUSAI K. & YAMANAKA T. (1973a) Cytochrome *c* (553, *Chlorobium thiosulfatophilum*) is a sulphide-cytochrome *c* reductase. *FEBS Lett.*, **34**, 235–7.
KUSAI K. & YAMANAKA T. (1973b) The oxidation mechanisms of thiosulphate and sulphide in *Chlorobium thiosulfatophilum*: roles of cytochrome *c*-551 and cytochrome *c*-553. *Biochim. Biophys. Acta*, **325**, 304–14.
LIPPERT K. D. (1967) Die Verwertung von molekularem Wasserstoff durch *Chlorobium thiosulfatophilum*. Thesis. University of Göttingen.
LIPPERT K. D. & PFENNIG N. (1969) Die Verwertung vom molekularem Wasserstoff durch *Chlorobium thiosulfatophilum*. *Arch. Mikrobiol.*, **65**, 29–47.
LLAMA M. J., SERRA J. L., RAS K. K. & HALL D. O. (1981) Isolation of two hydrogenase activities in *Chromatium*. *Eur. J. Biochem.*, **114**, 89–96.
MADIGAN M. T. & BROCK T. D. (1975) Photosynthetic sulfide oxidation by *Chloroflexus aurantiacus*, a filamentous, photosynthetic, gliding bacterium. *J. Bact.*, **122**, 782–4.
MADIGAN M. T. & GEST H. (1979) Growth of the photosynthetic bacterium *Rhodopseudomonas capsulata* chemoautotrophically in darkness with H_2 as the energy source. *J. Bact.*, **137**, 524–30.
NICHOLSON G. L. & SCHMIDT G. L. (1971) Structure of the *Chromatium* sulfur particle and its protein membrane. *J. Bact.*, **105**, 1142–8.
NIEL C. B. VAN (1963) A brief survey of the photosynthetic bacteria. In *Bacterial Photosynthesis* (Ed. by H. Gest, A. San Pietro & L. P. Vernon), pp. 459–67. Antioch Press, Yellow Springs.
OREN A. & PADAN E. (1978) Induction of anaerobic, photoautotrophic growth in the cyanobacterium *Oscillatoria limnetica*. *J. Bact.*, **133**, 558–63.
OREN A., PADAN E. & AVRON M. (1977) Quantum yields for oxygenic and anoxygenic photosynthesis in the cyanobacterium *Oscillatoria limnetica*. *Proc. natn. Acad. Sci. USA*, **74**, 2152–6.
OREN A. & SHILO M. (1979) Anaerobic heterotrophic dark metabolism in the cyanobacterium *Oscillatoria limnetica*: sulfur respiration and lactate fermentation. *Arch. Microbiol.*, **122**, 77–84.
ORMEROD J. G. & GEST H. (1962) Symposium on Metabolism of Inorganic Compounds. IV. Hydrogen photosynthesis and alternative metabolic pathways in photosynthetic bacteria. *Bact. Rev.*, **26**, 51–66.
PADAN E. (1979a) Impact of facultatively anaerobic photoautotrophic metabolism on ecology of cyanobacteria (blue-green algae). *Adv. Microb. Ecol.*, **3**, 1–48.
PADAN E. (1979b) Facultative anoxygenic photosynthesis in cyanobacteria. *Ann. Rev. Plant. Physiol.*, **30**, 27–40.
PAUL F., COLBEAU A. & VIGNAIS P. M. (1979) Phosphorylation coupled to H_2 oxidation by chromatophores from *Rhodopseudomonas capsulata*. *FEBS Lett.*, **106**, 29–33.
PESCHEK G. A. (1978) Reduced sulfur and nitrogen compounds and molecular hydrogen as electron donors for anaerobic CO_2 photoreduction in *Anacystis nidulans*. *Arch. Microbiol.*, **19**, 313–22.

PESCHEK G. A. (1979a) Anaerobic hydrogenase activity in *Anacystis nidulans*. H_2-dependent photoreduction and related reactions. *Biochim. Biophys. Acta*, **548**, 187–202.
PETUSHKOVA YU. P. & IVANOVSKII R. N. (1977) Enzymes participating in thiosulfate metabolism in *Thiocapsa roseopersicina* during its growth under various conditions. *Microbiology*, **45**, 822–7.
PFENNIG N. (1977) Phototrophic green and purple bacteria: a comparative systematic survey. *Ann. Rev. Microbiol.*, **31**, 275–90.
PFENNIG N. & SIEFERT E. (1977) Metabolism of C_1-compounds by *Rhodopseudomonas acidophila*. In *Abstracts of the 2nd International Symposium on Microbial Growth on C_1-compounds* (Ed. by G. K. Skryabin, M. V. Ivanov, E. N. Kondratieva, G. A. Zavarzin, Yu. A. Trotsenko & A. I. Nesterov), pp. 145–6. USSR Acad. Sciences, Puschino.
QUAYLE, J. R. & PFENNIG N. (1975) Utilization of methanol by Rhodospirillaceae. *Arch. Microbiol.*, **102**, 193–8.
RODOVA N. A. & PEDAN L. V. (1980) Metabolism of thiosulfate in *Rhodopseudomonas palustris*. *Microbiology*, **49**, 157–62.
ROELOFSEN P. A. (1935) On the photosynthesis of the Thiorhodaceae. Thesis. University of Utrecht.
ROLLS J. P. & LINDSTROM E. S. (1967a) Effect of thiosulfate on the photosynthetic growth of *Rhodopseudomonas palustris*. *J. Bact.*, **94**, 860–66.
ROLLS J. P. & LINDSTROM E. S. (1967b) Induction of a thiosulfate-oxidizing enzyme in *Rhodopseudomonas palustris*. *J. Bact.*, **94**, 784–5.
SAHM H., COX R. B. & QUAYLE J. R. (1976) Metabolism of methanol by *Rhodopseudomonas acidophila*. *J. gen. Microbiol.*, **94**, 313–22.
SCHEDEL M. (1978) Untersuchungen zur anaeroben Oxidation reduzierter Schwefelverbindungen durch *Thiobacillus denitrificans*, *Chromatium vinosum* und *Chlorobium limicola*. Thesis. University of Bonn.
SCHEDEL M., VANSELOW M. & TRÜPER H. G. (1979) Siroheme sulfite reductase isolated from *Chromatium vinosum*. Purification and investigation of some of its molecular and catalytic properties. *Arch. Microbiol.*, **121**, 29–36.
SCHMIDT G. L., NICHOLSON G. L. & KAMEN M. D. (1971) Composition of the sulfur particle of *Chromatium vinosum* strain D. *J. Bact.*, **105**, 1137–41.
SCHMITT W., SCHLEIFER G. & KNOBLOCH K. (1981) The enzymatic system thiosulfate: cytochrome *c* oxidoreductase from photolithoautotrophically grown *Chromatium vinosum*. *Arch. Microbiol.*, **130**, 328–33.
SCHNEIDER K. & SCHLEGEL H. G. (1977) Localization and stability of hydrogenases from aerobic hydrogen bacteria. *Arch. Microbiol.*, **112**, 229–38.
SCHWENN J. D. & BIERE M. (1979) APS-reductase activity in the chromatophores of *Chromatium vinosum* strain D. *FEMS Microbiol. Lett.*, **6**, 19–22.
SIEFERT E. & PFENNIG (1979) Chemoautotrophic growth of *Rhodopseudomonas* species with hydrogen and chemotrophic utilization of methanol and formate. *Arch. Microbiol.*, **122**, 177–82.
SLATER J. H. & MORRIS I. (1973) Photosynthetic carbon dioxide assimilation by *Rhodospirillum rubrum*. *Arch. Mikrobiol.*, **88**, 213–23.
SMITH A. J. (1966) The role of tetrathionate in the oxidation of thiosulphate by *Chromatium* sp. strain D. *J. gen. Microbiol.*, **42**, 371–80.
STEINMETZ M. A. & FISCHER U. (1981) Cytochromes of the non-thiosulfate-utilizing green sulfur bacterium *Chlorobium limicola*. *Arch. Microbiol.*, **130**, 31–37.
STOKES J. E. & HOARE D. S. (1969) Reductive pentose phosphate cycle and formate assimilation in *Rhodopseudomonas palustris*. *J. Bact.*, **100**, 890–4.
STOPPANI A. O. M., FULLER R. C. & CALVIN M. (1955) Carbon dioxide fixation by *Rhodopseudomonas capsulatus*. *J. Bact.*, **69**, 491–501.
STROHL W. R. & LARKIN J. M. (1978) Enumeration, isolation and characterization of *Beggiatoa* from freshwater sediments. *Appl. Environm. Microbiol.*, **36**, 755–70.

STROHL W. R., HOWARD K. S. & LARKIN J. M. (1982) Ultrastructure of *Beggiatoa alba* strain 15LD. *J. gen. Microbiol.*, **128**, 73–84.
THAUER R. K., JUNGERMANN K. & DECKER K. (1977) Energy conservation in chemotrophic anaerobic bacteria. *Bact. Rev.*, **41**, 100–80.
THIELE H. H. (1968) Sulfur metabolism in Thiorhodaceae. V. Enzymes of sulfur metabolism in *Thiocapsa floridana* and *Chromatium* species. *Antonie van Leeuwenhoek*, **34**, 350–6.
TIMMER-TEN HOOR A. (1976) Energetic aspects of the metabolism of reduced sulphur compounds in *Thiobacillus denitrificans*. *Antonie van Leeuwenhoek*, **42**, 483–92.
TRÜPER H. G. (1978) Sulfur metabolism. In *The Photosynthetic Bacteria* (Ed. by R. K. Clayton & W. R. Sistrom), pp. 677–90. Plenum Press, New York.
TRÜPER H. G. & PECK H. D. (1970) Formation of adenylylsulfate in phototrophic bacteria. *Arch. Mikrobiol.*, **73**, 125–42.
TRÜPER H. G. & ROGERS L. A. (1971) Purification and properties of adenylyl sulfate reductase from the phototrophic sulfur bacterium *Thiocapsa roseopersicina*. *J. Bact.*, **108**, 1112–21.
UFFEN R. L. (1976) Anaerobic growth of a *Rhodopseudomonas* species in the dark with carbon monoxide as sole carbon and energy substrate. *Proc. natn. Acad. Sci. USA*, **73**, 3298–302.
UFFEN R. L. (1981) Metabolism of carbon monoxide. *Enzyme Microb. Technol.*, **3**, 197–206.
UTKILEN H. C. (1976) Thiosulfate as electron donor in the blue-green alga *Anacystis nidulans*. *J. gen. Microbiol.*, **95**, 177–80.
WEAVER P. F., WALL J. D. & GEST H. (1975) Characterization of *Rhodopseudomonas capsulata*. *Arch. Microbiol.*, **105**, 207–16.
WERTLIEB D. & VISHNIAC W. (1967) Methane utilization by a strain of *Rhodopseudomonas gelatinosa*. *J. Bact.*, **93**, 1722–4.
WINFREY M. R., NELSON D. R., KLEVICKIS S. C. & ZEIKUS J. G. (1977) Association of hydrogen metabolism with methanogenesis in Lake Mendota sediments. *Appl. Environm. Microbiol.*, **33**, 312–18.
YOCH D. C. & LINDSTROM E. S. (1969) Nicotinamide adenine dinucleotide-dependent formate dehydrogenase from *Rhodopseudomonas palustris*. *Arch. Mikrobiol.*, **67**, 182–8.

Chapter 6. Essential Aspects of Carbon Metabolism

J. G. ORMEROD AND R. SIREVÅG

6.1 INTRODUCTION

As far as is known, all anoxygenic phototrophic bacteria can grow autotrophically, i.e. with CO_2 as sole carbon source and with an inorganic electron donor. Likewise, they can all utilize organic carbon for growth. The extent of these two nutritional abilities varies, some phototrophic bacteria being primarily autotrophic, e.g. the green sulphur bacteria, others, like the purple non-sulphur bacteria, primarily heterotrophic. It seems likely, however, that in many situations in nature the phototrophic bacteria utilize organic compounds, CO_2 and inorganic electron donors simultaneously.

Assimilation means entry of substrate carbon into the common pools of intermediates that are the starting points for biosynthesis of the major constituents of cell material. These key intermediates include acetyl CoA, pyruvate, phosphoenolpyruvate, oxaloacetate, α-oxoglutarate, succinyl CoA, and triose-, pentose-, and hexose-phosphates. Of course, in phototrophic bacteria, as in many other bacteria, these intermediates are largely interconvertible. Otherwise there would be no growth on single organic compounds or CO_2.

The ability to use light energy for growth under anaerobic conditions has several far-reaching physiological consequences. One is that organic carbon sources and inorganic electron donors need not be utilized for ATP production, and therefore the yield of cell mass as related to these substrates is much higher than in a chemotrophic anaerobe. Another consequence concerns the actual mechanisms of carbon assimilation since it appears that the types of assimilatory pathways operating in phototrophic bacteria may be determined by the nature of the primary products of the photosynthetic apparatus. Present knowledge (see Chapters 3 and 4) indicates that the redox potential of the primary reductant in purple bacteria is more positive than that in the green sulphur bacteria. This means that in purple bacteria, NAD cannot be reduced photosynthetically in a direct manner, but only by the more roundabout means of reversed electron transfer. On the other hand, the photosynthetic reductant in green sulphur bacteria is capable of directly reducing ferredoxin (Fd) which can in turn reduce NAD. This probably means that conditions within the illuminated cell of a green sulphur bacterium are more reducing than in a purple bacterium. Seen from this angle, it is not surprising that assimilation of CO_2 and organic compounds proceeds largely by different mechanisms in the two groups of organisms.

In this chapter, autotrophic CO_2 fixation is discussed first and thereafter the mechanisms involved in the assimilation of organic compounds.

6.2 CO_2 ASSIMILATION (IN THE ABSENCE OF ORGANIC SUBSTRATES)

One of the major points of interest with regard to assimilation of CO_2 in the phototrophic bacteria is the mechanism by which this takes place. In aerobic phototrophic organisms like higher plants, algae and bluegreen bacteria, the mechanism CO_2-fixation is the Calvin cycle shown in Fig. 6.1. This cycle is also widespread among organisms with anoxygenic photosynthesis, notably the purple bacteria. However, its presence in Chlorobiaceae is dubious and has been a matter of controversy. For members of this group, e.g. *Chlorobium*, an alternative mechanism of CO_2-fixation, the reductive tricarboxylic acid cycle (Fig. 6.2) has been proposed (Evans *et al.*, 1966) and the concept of this cycle as a mechanism of CO_2-fixation is now generally accepted, despite recent assertions to the contrary (Fuller, 1978; McFadden, 1978).

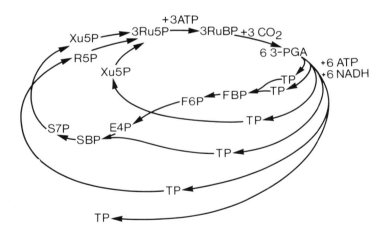

Fig. 6.1 The Calvin cycle. E4P, erythrose-4-phosphate; F6P, fructose-6-phosphate; FBP, fructose-1,6-bisphosphate; 3-PGA, 3-phosphoglycerate; R5P, ribose-5-phosphate; Ru5P, ribulose-5-phosphate; RuBP, ribulose-1,5-bisphosphate; S7P, sedoheptulose-7-phosphate; SBP, Sedoheptulose-1,7-bisphosphate; TP, triosephosphate; Xu5P, xylulose-5-phosphate.

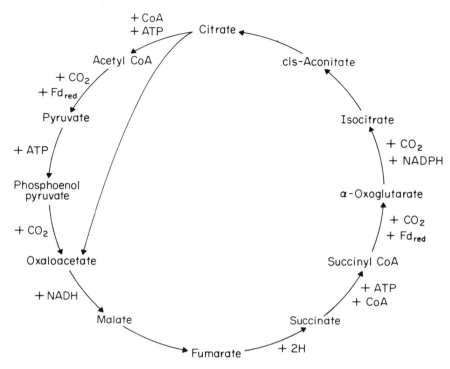

Fig. 6.2 The reductive tricarboxylic acid cycle of green sulphur bacteria. Fd_{red}, reduced ferredoxin.

6.2.1 The Calvin cycle

The reactions of the Calvin cycle are outlined in Fig. 6.1. Of the several steps involved, those catalysed by phosphoribulokinase and ribulose 1,5-bisphosphate (RuBP) carboxylase are commonly considered key reactions of the cycle and their presence is usually taken as an indication of the presence of the complete Calvin cycle. Of the two key enzymes, RuBP carboxylase which catalyses the fixation of CO_2 into 3-phosphoglyceric acid (3-PGA) as shown below, has been subject to the most extensive examination, and its molecular and chemical properties are now fairly well understood. (Reaction 6.1, page 103.) The 3-PGA formed in this reaction is further metabolized to sugar phosphates by the subsequent reactions of the cycle.

In addition to the above reaction, RuBP carboxylase catalyses the oxygenolytic cleavage of RuBP in the presence of O_2 to yield 3-PGA and phosphoglycollate. (Reaction 6.2, page 103.) In green plants, phosphoglycollate formed in this reaction is converted to glycollate which is used as a substrate for photorespiration.

Reaction 6.1

$$CO_2 + \text{RuBP} \xrightarrow{Mg^{2+}} \text{unstable intermediate} \longrightarrow \text{3-PGA}$$

where RuBP is:

```
CH₂O—P
|
C=O
|
H—COH
|
H—COH
|
CH₂O—P
```

unstable intermediate:

```
CH₂O—P
|
C(OH)—COOH
|
C=O
|
H—COH
|
CH₂O—P
```

3-PGA (two molecules):

```
CH₂O—P              COOH
|                   |
H—COH               H—COH
|                   |
COOH                CH₂O—P
```

Reaction 6.2

$$O_2 + \text{RuBP} \longrightarrow \text{P-glycollate} + \text{3-PGA}$$

P-glycollate:
```
CH₂O—P
|
COOH
```

3-PGA:
```
COOH
|
H—COH
|
CH₂O—P
```

The activity of RuBP carboxylase is usually assayed in cell-free extracts by measuring the RuBP-dependent incorporation of $^{14}CO_2$ into acid-stable material. Using this method the enzyme has been found in several purple sulphur and non-sulphur bacteria and is probably universally present in these (see Kondratieva, 1979 for review). The presence of the enzyme has also been reported in the green bacteria *Chlorobium* (Smillie et al., 1962; Tabita et al., 1974) and *Chloroflexus* (Madigan & Brock, 1977; Sirevåg & Castenholz, 1979).

In the case of *Chlorobium*, conflicting data exist regarding the presence of RuBP carboxylase. Low levels of the enzyme were first reported by Smillie et al. (1962) using a spectrophotometric method. Later, Tabita et al. (1974) reported that low levels of RuBP carboxylase could readily be detected in this organism by the radioassay both in crude cell-free extracts and in purified enzyme fractions obtained by sucrose density gradient centrifugation. Several characteristics of the enzyme, including its stability, molecular weight and quaternary structure were described.

However, in contrast to these findings are reports from other workers (Buchanan et al., 1972; Sirevåg, 1974; Takabe & Akazawa, 1977; Quandt, 1978) who have been unable to detect either RuBP carboxylase or phosphoribulokinase in extracts of *Chlorobium*. In the work reported by Takabe & Akazawa (1977), the possible oxygenolytic activity of RuBP-carboxylase was also investigated, but no trace of glycollate was found when *Chlorobium* cells were exposed to O_2.

In a critical examination of the problem, Buchanan & Sirevåg (1976) used exactly the same procedure as Tabita et al. (1974), and like these authors, observed a slow, RuBP-dependent fixation of $^{14}CO_2$. However, it was concluded that the fixation could not be due to RuBP carboxylase since the stable radioactive reaction product was a mixture of malate, asparate and glutamate and *not* 3-PGA as would be expected if RuBP carboxylase were active.

RuBP carboxylase is a relatively large protein whose quaternary structure has been well characterized in a large number of organisms (see McFadden, 1978 for a review). The enzyme from different sources varies with regard to molecular weight and subunit structure: it may contain subunits of two different types, a large (L) type, of molecular weight 55 000 and a small (S) type of molecular weight 15 000. Generally RuBP carboxylase from higher plants and algae comprises 8 subunits of each type and is designated L_8S_8, whereas more variation in the structure is found among phototrophic bacteria.

Except for *Rs. rubrum* in which RuBP carboxylase is relatively small and consists of only two large subunits (L_2S_0), it appears that all strains of Rhodospirillaceae examined synthesize the plant-type RuBP carboxylase, L_8S_8 (Tabita, 1981). However, some species, like *Rp. palustris* and *Rp. sphaeroides* synthesize in addition an RuBP carboxylase of the type: L_6S_0 (Anderson et al., 1968; Gibson & Tabita, 1977; Tabita, 1981). Detailed immunological and structural analysis of isolated large subunits from the two enzymes has revealed that they are distinct, and coded for by different genes, a fact which has led to the suggestion that plasmid DNA might be involved in the coding (Tabita, 1981).

The important implication of the fact that some active RuBP carboxylases consist of subunits of the large type only, is that the catalytic site of the enzyme must be localized on this subunit (Tabita & McFadden, 1974).

The molecular weight and subunit structure of the putative RuBP carboxylase from *Chlorobium* have also been determined: L_6S_0 (Tabita et al., 1974), but serious doubt has later been raised as to whether the protein studied by these authors in fact was this enzyme (Buchanan & Sirevåg, 1976).

Short-term labelling experiments

In addition to enzymic studies one can gain information about the mechanism of CO_2 fixation from short-term labelling experiments in which the cells are exposed to $^{14}CO_2$ for various periods of time. The radioactive products in the suspensions are then analysed and identified. With cells like algae, which have a functional Calvin cycle, a characteristic pattern of distribution of the radioisotope is obtained. After very short exposures (1 s), most of the radioactivity is recovered in the first stable product of CO_2-fixation, 3-PGA. The relative amount of radioactivity in this compound decreases with time as other intermediates in the cycle are formed from it. Thus, when the percentage of total radioactivity in 3-PGA is plotted against time, a characteristic negative slope is obtained. However, this type of experiment can be misleading, the difficulty being to prove that the negative slope actually begins at zero time and is not preceded by an initial, positive slope.

The kinetics of short-term labelling with $^{14}CO_2$ have been studied in *Chromatium* (Fuller *et al.*, 1961), *Rp. sphaeroides* (Stoppani *et al.*, 1952) and in *Rs. rubrum* (Glover *et al.*, 1952; Anderson & Fuller, 1967) and are consistent with the operation of the Calvin cycle in these organisms.

However, the labelling pattern with *Chlorobium* is quite different and far more complicated. In this case, various organic acids (amino acids, α-oxo acids, carboxylic acids) rather than phosphate esters are the early radioactive products formed from $^{14}CO_2$ (Evans *et al.*, 1966; Buchanan *et al.*, 1972; Sirevåg, 1974), results which comprise further evidence *against* the operation of the Calvin cycle in this organism.

Carbon isotope fractionation

The finding that RuBP carboxylase, in contrast to other carboxylases, shows a much greater preference for $^{12}CO_2$ than for $^{13}CO_2$ (Whelan *et al.*, 1973) can be used to distinguish organisms which use different carboxylation reactions for initial CO_2-fixation by comparing their carbon isotope composition after autotrophic growth.

Sirevåg *et al.* (1977) performed this kind of analysis with representatives of the different groups of phototrophic bacteria grown autotrophically. It was found that the purple bacteria *Chromatium* and *Rs. rubrum* had isotope discrimination characteristics similar to those of other organisms known to have the Calvin cycle, whereas *Chlorobium* was radically different in this respect. These results were supported by similar findings by Quandt *et al.* (1977) and strongly indicate that other carboxylation enzymes than RuBP carboxylase contribute to growth in *Chlorobium*.

Regulation of activity of the Calvin cycle enzymes

In addition to an adequate supply of CO_2, functioning of the Calvin cycle depends on ATP, reducing power and a suitable ionic environment. Consequently, factors like light, CO_2 concentration and various ions influence its activity.

In chloroplasts and cyanobacteria certain of the enzymes of the Calvin cycle are activated by light through a mechanism involving thioredoxin and ferredoxin-thioredoxin reductase (Wolosiuk & Buchanan, 1977). The cycle enzymes activated by this mechanism are fructose bisphosphatase, sedoheptulose bisphosphatase and a phosphoribulokinase. It has also been shown in chloroplasts (Walker & Robertson, 1978) that the concentration of the two acceptors of ATP in the cycle, ribulose 5-phosphate (Ru5P) and 3-PGA, regulate the total activity of the cycle in such a way that high levels of 3-PGA and low levels of Ru5P are necessary for high rates of CO_2-fixation. Whether or not regulatory mechanisms like these operate in purple bacteria is at present unknown, however. Intracellular Mg^{2+} concentration is probably an important universal regulator of RuBP carboxylase since the enzyme is activated by combination with Mg^{2+} and CO_2 (Badger, 1980).

Regulation of biosynthesis of the Calvin cycle enzymes

The nutritional versatily of some phototrophic bacteria makes them well-suited for the study of regulation of synthesis of enzymes concerned with CO_2 assimilation.

This approach was used by Anderson & Fuller (1967a; b; c) who measured enzyme levels in cells of *Rs. rubrum* grown photoautotrophically and photoheterotrophically as well as aerobically in the dark. They found that growth on malate and to a lesser extent acetate, decreased the levels of RuBP carboxylase and fructose bisphosphatase as compared to those in autotrophically grown cells.

On the other hand, Tabita & McFadden (1974) found that when *Rs. rubrum* was grown on a reduced organic substrate like butyrate, the synthesis of RuBP carboxylase was derepressed and levels of the enzyme comparable to those in autotrophic cells were obtained. Later it was reported that the derepression did not occur until the onset of the stationary phase (see Section 6.3.2) and further, that the other key enzyme of the Calvin cycle, phosphoribulokinase, appears to be controlled coordinately with RuBP carboxylase in such cultures, suggesting that the genes encoding these two enzymes might be located on the same operon (Tabita, 1981).

6.2.2 The reductive tricarboxylic acid cycle

The reductive tricarboxylic acid cycle as shown in Fig. 6.2 was proposed by Evans *et al.* in 1966. Their proposal was based on the discovery in phototrophic bacteria and other anaerobes of two new ferredoxin dependent carboxylation reactions:

$$\text{acetyl-CoA} + CO_2 + Fd_{red} \longrightarrow \text{pyruvate} + \text{CoA} + Fd_{ox}$$
$$\text{succinyl-CoA} + CO_2 + Fd_{red} \longrightarrow \alpha\text{-oxoglutarate} + \text{CoA} + Fd_{ox}$$

catalysed by pyruvate synthase and α-oxoglutarate synthase respectively (Evans & Buchanan, 1965; Buchanan & Evans, 1965; Bachofen, 1964). The cycle involves reversal of the reactions in the tricarboxylic acid cycle and results in the net synthesis of one molecule of acetate from two of carbon dioxide. Acetyl-CoA is partly removed for biosynthesis or recycled for anaplerotic purposes.

Originally, it was suggested that the main function of the cycle was in the formation of precursors for amino acids, lipids and porphyrins since it was assumed that the Calvin cycle accounted for carbohydrate formation (Evans et al., 1966). However, the accumulating evidence (see above) indicates that Chlorobium lacks the Calvin cycle and several other lines of evidence now suggest that the reductive tricarboxylic acid cycle is the sole mechanism of CO_2 fixation in Chlorobium.

Experimental evidence

Evans et al. (1966) reported that all the enzymes of the cycle were present in cell-free extracts of Chlorobium. Further support for the cycle was their finding that $^{14}CO_2$ was rapidly incorporated into amino acids, particularly glutamate.

A major weakness of their data is that the activity of several of the enzymes measured was too low to account for CO_2 fixation in whole cells during growth. Further, their results with $^{14}CO_2$ can easily be explained by assuming that the two ferredoxin linked carboxylations are independent reactions, not forming part of a cycle. One of the key reactions of the reductive tricarboxylic acid cycle is the cleavage of citrate by citrate lyase into acetyl CoA and oxaloacetate. This enzyme was measured by an indirect method by Evans et al. (1966), but could not be detected by Beuscher & Gottschalk (1972), a fact which has been considered a serious objection to the cycle as outlined originally. However, very recently Ivanovsky et al. (1980) reported the presence of an ATP-dependent citrate lyase in cell-free extracts of Chlorobium, a finding which has now been fully confirmed in our laboratory.

Evidence for the operation of the reductive tricarboxylic acid cycle as the *sole* mechanism of CO_2-fixation was obtained by Sirevåg and Ormerod (1970a; b), using washed cell suspensions of Chlorobium in respirometers. Most of the CO_2 fixed was accounted for by the formation of various α-oxoacids and polyglucose. Fluoroacetate, which is known to inhibit the citric acid cycle, was employed in these studies to inhibit the reductive tricarboxylic acid cycle. When this compound was added to the washed suspensions, CO_2-fixation was strongly inhibited and an increased accumulation of α-oxoglutarate was observed. At the same time, formation of other α-oxoacids and polyglucose was inhibited, suggesting that glucose formation is dependent on the operation of the reductive tricarboxylic acid cycle in Chlorobium. Another observation with washed cell suspensions which supports this point, is that addition of acetate caused an increase in exactly the same products as those formed from CO_2 alone, namely α-oxoacids and polyglucose. Since acetate is easily convertible into acetyl CoA, which is an intermediate of the reductive tricarboxylic acid cycle, this indicates that the cycle contributes not only to the formation of amino acids, lipids and porphyrins but also to biosynthesis of carbohydrate.

Several other lines of evidence now support this conclusion: Using acetate and CO_2 labelled with ^{14}C in various combinations, Sirevåg (1975) showed that acetate after carboxylation to pyruvate is converted to polyglucose in *Chlorobium*. Degradation of the glucose thus formed, indicated that a mechanism similar to a reversed glycolysis was functioning. Essentially the same results were also obtained in experiments where NMR was used to analyse glucose formed from $^{13}CO_2$ and acetate in growing cultures of *Chlorobium* (Paalme *et al.*, 1982). Similarly, Fuchs *et al.* (1980a) in studying the metabolism of specifically labelled ^{14}C-pyruvate into amino acids and glucose in growing cells of *Chlorobium* showed that hexose phosphate synthesis proceeds solely *via* pyruvate, thus excluding any role of the Calvin cycle in carbohydrate formation.

The formation of pyruvate from acetyl-CoA is an important step in the reductive tricarboxylic acid cycle. The enzyme involved, pyruvate synthase, is inhibited by glyoxylate (Thauer *et al.*, 1970) and in keeping with this, Quandt *et al.* (1978) showed that glyoxylate completely inhibits growth, pyruvate synthase and CO_2-fixation in *Chlorobium*. More recently, (Ormerod, 1980), it was shown that in cultures of *Chlorobium* inhibited by glyoxylate, acetate accumulates at a rate approximating that of CO_2-fixation in uninhibited cultures at the same stage of growth. This means that acetate (as acetyl-CoA) is a true intermediate in CO_2 fixation in *Chlorobium*.

The evidence from short-term labelling experiments with $^{14}CO_2$ in *Chlorobium* is complex owing to the activity of the numerous carboxylation reactions in this organism. In general it can be said that α-oxoacids of the cycle constitute the early labelled products. (Sirevåg, 1974), sugar phosphates appearing later (Buchanan *et al.*, 1972).

6.2.3 Concluding remarks

Today the presence of the reductive tricarboxylic acid cycle in *Chlorobium* is generally accepted and is not subject to controversy. However, what has been controversial is its significance and the idea of a photoautotrophic organism that lacks the Calvin cycle. From an evolutionary point of view this is not implausible. It is widely assumed that the earliest living organisms were strictly anaerobic chemoheterotrophs and that phototrophic organisms evolved from these. It is therefore not unlikely that the green sulphur bacteria are relics from a time when the conditions on earth were highly reductive and thus would facilitate the operation of a mechanism like the reductive tricarboxylic acid cycle with its requirement for an environment of low redox potential.

According to this line of reasoning the Calvin cycle would be a later development. It has the majority of its enzymes in common with the oxidative pentose phosphate cycle, reactions of which are needed in most cells to provide pentose phosphate for nucleotide synthesis. Thus, the Calvin cycle may have evolved from the oxidative pentose phosphate cycle by addition of RuBP carboxylase and phosphoribulokinase (Quayle & Ferenci, 1978). The Calvin cycle has almost without exception been found only in aerobic or aerotolerant organisms and the oxygenase activity of RuBP carboxylase may well reflect a specific aerobic function. Indeed, the possibility that the original function of the Calvin cycle was to consume oxygen by the RuBP oxygenase reaction must be considered seriously in view of the widespread occurrence of photorespiration. Accordingly, autotrophic organisms possessing the Calvin cycle may be viewed as more recently evolved than the green sulphur bacteria, which are almost unique among the Eubacteria in being both strictly anaerobic and autotrophic.

6.3 HOW ORGANIC COMPOUNDS ARE METABOLIZED

As is apparent from the previous section, the green sulphur bacteria are in many ways metabolically distinct from the purple bacteria and the two groups are therefore dealt with separately in this section. In addition a brief subsection on carbon metabolism in *Chloroflexus* is included.

6.3.1 Green sulphur bacteria

Assimilatory mechanisms

The CO_2-fixing mechanism of green sulphur bacteria, the reductive tricarboxylic acid cycle (Section 6.2.2) provides the core of reactions for assimilation of organic substrates. This cycle can only function in the reductive direction since α-oxoglutarate dehydrogenase is lacking in *Chlorobium* (Beatty & Gest, 1981). Therefore, no CO_2 or reducing power are available from the cycle for metabolic conversions. Consequently, added organic molecules can only be assimilated provided that substrate amounts of CO_2 and an inorganic electron donor such as H_2S are present. These requirements for assimilation were first reported by Sadler & Stanier (1960) and they distinguish the green sulphur bacteria from all other phototrophic bacteria.

Once a substrate has entered the reductive tricarboxylic acid cycle, it becomes converted into intermediates that are the starting points for biosynthesis of fatty acids (from acetyl CoA), almost all the amino acids (from pyruvate, oxaloacetate and α-oxoglutarate) and carbohydrates (from phosphoenol pyruvate [PEP]). In the case of carbohydrates, it has been shown (Sirevåg, 1975) that glucose is formed *via* the Embden-Meyerhof route. Presumably tetrose- and pentose-phosphates are produced by enzymes of the oxidative pentose phosphate mechanism.

The organic substrates that can be assimilated by the green sulphur bacteria are acetate, propionate, pyruvate and, in some strains, lactate, glutamate and glucose. Of these, acetate is by far the best substrate; its addition to an autotrophically growing culture almost doubles the growth rate and yield. Evidence indicating that acetate is assimilated *via* the reductive tricarboxylic acid cycle was in fact available (Hoare & Gibson, 1964) before this mechanism had been proposed. In this pioneering work amino acids from the protein of *Chlorobium* cells grown on labelled acetate and CO_2 were degraded. The results are in keeping with assimilation of acetate *via* the reductive tricarboxylic acid cycle, provided that recycling is taken into account.

The growth of *Chlorobium* is also stimulated by propionate, which enters the cycle by way of succinyl-CoA formed by activation and carboxylation of the propionate. This important mechanism was discovered by Larsen (1951) working in van Niel's laboratory and this was the first demonstration of the reaction (now known to be universal) in a biological system. More recently, Fuchs *et al.* (1980b) fed labelled propionate and labelled pyruvate to *Chlorobium* cultures and isolated alanine, aspartate and glutamate from the cell protein. Partial degradation of these amino acids revealed labelling patterns that were in full agreement with assimilation of both propionate and pyruvate by the reactions of the reductive tricarboxylic acid cycle.

Regulation of carbon metabolism in green sulphur bacteria

There are no data concerning the regulation of enzyme activity in *Chlorobium*. In contrast to the Calvin cycle enzymes in purple bacteria (Section 6.1), the synthesis of the enzymes of the reductive tricarboxylic acid cycle is apparently not subject to extensive regulation. The evidence for this is as follows:

1. The specific activity of aconitate hydratase and isocitrate dehydrogenase is the same in cells grown autotrophically as in acetate-grown cells (Sirevåg, 1974);
2. When acetate is added to an autotrophically growing culture there is an immediate increase in growth rate, indicating that the constitutive levels of the cycle enzymes involved are high enough to support the higher rate of growth with acetate.

Carbon metabolism of *Chlorobium* is now fairly well understood, and there are reasons for believing that the green sulphur bacteria as a group are metabolically homogeneous (Sirevåg & Ormerod, 1970a).

6.3.2 The green gliding bacteria

These bacteria grow best on organic substrates and are also able to grow well aerobically in the dark (Madigan et al., 1974). In these respects they resemble the purple non-sulphur bacteria and there is evidence that they utilize the forward citric acid cycle for metabolizing organic substrates in both the light and the dark (Sirevåg & Castenholz, 1979). In addition, it has been shown that when *Chloroflexus* grows with acetate as sole organic carbon source, the glyoxylate cycle is involved in acetate utilization (Løken & Sirevåg, 1982).

6.3.3 Purple bacteria

Compared with the green sulphur bacteria, the purple bacteria are metabolically much more heterogeneous, and some, like species of the genera *Rhodospirillum* and *Rhodopseudomonas* are quite versatile. These are the species that have been most intensively investigated and they have been shown to utilize a whole range of simple organic acids, alcohols and other compounds. Some can grow on aromatic compounds, others on C_1-compounds. The purple sulphur bacteria on the other hand, are on the whole more limited in their ability to use organic substrates.

The photoassimilation of organic substrates is very efficient. If the oxidation state of a substrate is higher than that of cell material ($C_5H_8O_2N$; van Gemerden, 1968a), some CO_2 is produced; if the substrate is more reduced than cell material, CO_2 is fixed. Usually no other products than CO_2 can be detected in the medium under optimum growth conditions.

Because of their heterogeneity it is difficult to generalize on the carbon metabolism of the purple bacteria although one thing seems to be clear: reactions of the oxidative citric acid cycle lie at the centre of the assimilation processes of most of the organic substrates. There is no evidence for the operation of the complete reductive tricarboxylic acid cycle in purple bacteria but segments of the citric acid cycle may be proceeding in reverse in the case of some substrates.

Three examples of photoassimilation of organic substrates by *Rhodospirillum rubrum* are described below. These have been chosen in order to emphasize the complexity of the metabolism of these bacteria.

The first example concerns the assimilation of succinate. The empirical formula of this substrate is $C_4H_6O_4$, which is more oxidized than cell material. Accordingly, assimilation results in the formation of CO_2, about 0.7 mol/mol succinate. However, measurements with $^{14}CO_2$ show that there is a concealed simultaneous fixation of CO_2, amounting to about 0.5 mol/mol succinate (Ormerod, 1956).

There are several lines of evidence (Elsden & Ormerod, 1956; Knight, 1962; Evans, 1965) to suggest that the following sequence of reactions is involved in the photometabolism of succinate:

$$\text{succinate} \longrightarrow \text{fumarate} \longrightarrow \text{malate} \longrightarrow \text{oxaloacetate} \longrightarrow \text{PEP}$$

In addition, growing cells must form α-oxoglutarate, which is probably the most important intermediate in the assimilation process. In *Rs. rubrum*, α-oxoglutarate can in theory be formed *via* citrate or by the reductive carboxylation of succinyl CoA (catalysed by α-oxoglutarate synthase; Buchanan *et al.*, 1967). The latter mechanism seems most likely, since the results of investigations with fluoroacetate (Elsden & Ormerod, 1956) and of short-term labelling experiments (Knight, 1962) argue against the participation of citrate here. Also, Shigesada *et al.*, (1966) showed that in the presence of unlabelled succinate, $^{14}CO_2$ was incorporated into C_1 of glutamate, while 1,4-^{14}C succinate gave labelling in C_2 and C_5. Thus the conversion of succinate into central intermediates may be envisaged as follows:

$$\text{α-oxoglutarate} \swarrow \overset{\text{succinate}}{\searrow} \text{oxaloacetate} \rightarrow \text{PEP} \rightarrow \text{pyruvate} \rightarrow \text{acetylCoA}$$

In other words, one section of the tricarboxylic acid cycle is operating in the forward direction while another is operating in reverse.

The second example of assimilation in *Rs. rubrum* is that of acetate. This substrate is slightly more oxidized than cell material and its assimilation is therefore accompanied by the production of about 0.2 mol CO_2/mol. Of central importance in acetate metabolism is the mechanism by which C_4 acids are produced and thus far, three such mechanisms have been described in purple bacteria:

(i) Many of these bacteria can synthesize the enzymes of the glyoxylate cycle (Albers & Gottschalk, 1976);
(ii) In *Rp. gelatinosa*, C_4 acids are formed from acetate by a mechanism involving the serine-hydroxypyruvate pathway (Albers & Gottschalk, 1976);
(iii) In *Rs. rubrum*, there is good evidence (Cutinelli *et al.*, 1951; Buchanan *et al.*, 1967) that acetate is reductively carboxylated to pyruvate and the pyruvate converted to oxaloacetate.

However, in *Rs. rubrum* the situation with regard to C_4 acid synthesis from acetate is more complicated than the above would suggest. Thus, Hoare (1963) showed that oxaloacetate formed according to mechanism (iii) could not be an intermediate in the synthesis of glutamate from labelled acetate. Degradation of the labelled glutamate showed the following pattern:

$$\begin{array}{cccccc} & 1 & 2 & 3 & 4 & 5 \\ \text{glutamate} & \text{HOOC} . & \text{CHNH}_2 . & \text{CH}_2 . & \text{CH}_2 . & \text{COOH} \\ \text{origin} & CO_2 & \underbrace{\text{HOOC} \quad \text{CH}_3}_{\text{acetate}} & & \underbrace{\text{CH}_3 \quad \text{COOH}}_{\text{acetate}} & \end{array}$$

If the glutamate had been formed (*via* citrate) from oxaloacetate produced as described above (Cutinelli *et al.*, 1951), the origins of C_2 and C_3 in glutamate would have been reversed. Thus, in spite of the fact that all of the enzymes required for the formation of glutamate from acetate by reactions of the 'forward' citric acid cycle are present, this mechanism does not proceed further than oxaloacetate. The evidence indicates that instead glutamate is formed from two molecules of acetate or its derivative(s), which react 'back to back' to give a C-4 acid which is then converted by carboxylation into α-oxoglutarate. The details of the mechanism are not clear. Experimental data on how other purple bacteria form glutamate are completely lacking.

The third substrate to be considered here is butyrate ($C_4H_8O_2$) which is more reduced than cell material. There is no CO_2 production during the assimilation of this substrate, but rather a CO_2 fixation, amounting to about 0.4 mol/mol butyrate. It has generally been assumed that this CO_2 fixation occurs *via* the Calvin cycle. However, as shown by Tabita (1981), significant Calvin cycle activity does not appear in cultures before the onset of the stationary phase. A possible explanation for this result is that butyrate is first oxidized to acetyl CoA which is then reductively assimilated in the same manner as acetyl CoA formed from exogenous acetate (see above). The appearance of Calvin cycle activity at the end of growth may be connected with the conversion of poly-β-hydroxybutyrate (the reserve material formed from butyrate) into other more oxidized cell constituents, with consequent formation of reducing power for CO_2 fixation.

The three substrates discussed above are among those which have been most intensively investigated in purple bacteria. The large amount of research carried out in this field has resulted in surprisingly few unequivocal conclusions. The reason for this seems to be the great variety of metabolic pathways and regulatory mechanisms in purple bacteria. Against this background, the measuring of enzyme activities and identifying of labelled substances on chromatograms, although important, do not prove that a particular mechanism is operating; decisive proof requires the isolation and degradation of labelled intermediates, and since these operations are laborious and time-consuming, they are not often embarked upon. The use of mutants in this field would be expected to give worthwhile results.

We also possess some knowledge of the assimilation of a number of other substrates by these organisms. For example, some purple non-sulphur bacteria can grow phototrophically on aromatic compounds like benzoate. The process is strictly anaerobic and light-dependent (Dutton & Evans, 1969) and metabolism of the aromatic compound involves complete reduction of the benzene ring followed by opening to form pimelyl CoA (Whittle *et al.*, 1976).

All of the assimilatory processes considered above occur in cultures in which an ammonium salt is the nitrogen source. The metabolism of organic substrates may be very different in cells of purple bacteria grown with other nitrogen sources, such as N_2 or glutamate (Gest & Kamen, 1949). Such cells can form hydrogen during the assimilation of organic substrates and in the absence of nitrogen can convert the substrate completely into CO_2 and molecular hydrogen (Gest et al., 1962). The process is light-dependent, strictly anaerobic and involves nitrogenase and the citric acid cycle. Such total degradation of substrates to CO_2 and H_2 is a manifestation of extreme nitrogen starvation.

Finally, it should be mentioned that many purple non-sulphur bacteria can grow on organic substrates aerobically in the dark. Under these conditions, metabolism resembles that of typical aerobic chemoheterotrophs, with the citric acid cycle functioning as the main source of electrons and biosynthetic intermediates.

6.4 CARBON RESERVE MATERIALS AND ENDOGENOUS METABOLISM

In nature, the phototrophic bacteria are exposed to regular variations in energy supply (night/day) and this places a stringent requirement for the synthesis of reserve materials in daylight to serve for maintenance metabolism at night. During balanced growth, the phototrophic bacteria usually synthesize small amounts of reserve material (equivalent to a few percent of the dry weight), but under conditions of nitrogen starvation, the amounts formed may be very much greater.

There are two carbonaceous reserve materials formed in the phototrophic bacteria, namely glycogen and poly-β-hydroxybutyrate. The capacity to synthesize both of these polymers is found in representatives of all families of anoxygenic phototrophs except the green sulphur bacteria, which synthesize glycogen only.

6.4.1 Glycogen

Glycogen is a large polymer in which glucose residues are joined together by $\alpha(1, 4)$ glucosidic linkages into long chains which are branched, with $\alpha(1, 6)$ linkages at the branch points. The common precursor for glycogen synthesis in all organisms is glucose-1-phosphate which in prokaryotes is converted to ADP-glucose in a reaction catalysed by ADP-glucose phosphorylase:

$$ATP + \text{glucose-1-phosphate} \longrightarrow \text{ADP-glucose} + PP_i$$

(In eukaryotes, the corresponding reaction involves UTP)

Two additional enzymes are further required for glycogen synthesis in prokayotes. One, ADP-glucose: 1,4-glucan transferase, is specific for the formation of the $\alpha(1, 4)$ linkages:

$$\text{ADP-glucose} + (\text{glucosyl})_n \longrightarrow ADP + (\text{glucosyl})_{n+1}$$

The other enzyme is the 'branching' enzyme, which forms the $\alpha(1, 6)$ linkages by transferring part of an outer chain into the 6-position of a glucose residue further into the molecule.

Glycogen is deposited in prokaryotic and eukaryotic cells in the form of non membrane-bound granules with a rosette-like structure made up of smaller particles (Candy, 1980). When the deposition of glycogen in cells of *Chlorobium* was examined by electron microscopy of thin sections stained specifically for carbohydrate, results corresponding to this general picture were obtained (Sirevåg & Ormerod, 1977). The size of the granules in *Chlorobium* was limited to 30 nm and as the glycogen content of the cells increased, the number, rather than the size of the granules increased. The granules in *Rs. rubrum* are larger, however, and are visible in thin sections as light areas measuring 50–150 nm scattered throughout the cytoplasm (Cohen-Bazire & Kunisawa, 1963).

6.4.2 Poly β-hydroxybutyrate

This substance (PHB) is a linear polymer of D-β-hydroxybutyric acid:

$$-O-\underset{\underset{CH_3}{|}}{CH}.CH_2.CO-O-\underset{\underset{CH_3}{|}}{CH}.CH_2.CO-O-\underset{\underset{CH_3}{|}}{CH}.CH_2.CO-$$

The molecular weight of PHB may be as high as several hundred thousand (Lundgren *et al.*, 1965) and the intracellular granules of the polymer in *Rs. rubrum* measure up to about 400 nm and are surrounded by a thin, non-unit membrane (Boatman, 1964). The polymer has rather unusual chemical properties, being insoluble in ether but soluble in hot chloroform. Cultures which form PHB may acquire a chalky appearance.

The synthesis of PHB has been examined in *Rs. rubrum* (Moskowitz & Merrick, 1969). It does not involve acyl carrier protein and is believed to proceed as follows:

2 acetylCoA ⟶ acetoacetylCoA ⟶ L-β-hydroxybutyrylCoA
⟶ crotonylCoA ⟶ D-β-hydroxybutyrylCoA ⟶ PHB

Purple non-sulphur bacteria form PHB during phototrophic growth on organic substrates that are metabolized via acetyl CoA, for example acetate and butyrate (Stanier *et al.*, 1959). Washed suspensions of *Chloroflexus* also form PHB from acetate provided that accessory reducing power in the form of H_2 is supplied (Sirevåg & Castenholz, 1979).

6.4.3 Breakdown of reserve materials; endogeneous metabolism

Intracellular glycogen formed by purple non-sulphur bacteria in the light is fermented under anaerobic conditions in the dark to a mixture of volatile fatty acids (Kohlmiller & Gest, 1951). This process supplies ATP for maintenance purposes and is probably triggered by lowering of the energy charge (Atkinson, 1968) as the light fails, since it can be induced by adding inhibitors of photosynthetic phosphorylation to illuminated cells (Gest et al., 1962). In a similar manner, green sulphur bacteria also break down endogeneous glycogen in the dark to form volatile fatty acids plus succinate (Sirevåg & Ormerod, 1977).

In *Chromatium*, it was shown by van Gemerden (1968b) that glycogen formed from CO_2 during the oxidation of sulphide to elemental sulphur was converted into other cell constituents during subsequent oxidation of the sulphur to sulphate (see Chapter 5, Fig. 5.1). Furthermore, glycogen formed in the light becomes converted into PHB in the dark, elemental sulphur acting as electron acceptor and thereby becoming reduced to H_2S. This process provides ATP for maintenance purposes while at the same time most of the carbon is retained. Thus, we see in the purple sulphur bacteria a remarkable flexibility in the metabolism of reserve materials.

We know much less about the metabolism of PHB in these bacteria. The initial enzymic hydrolysis yields free β-hydroxybutyrate, not the CoA derivative. The β-hydroxybutyrate can be converted into acetyl CoA by the expenditure of energy but the detailed fate of PHB under anaerobic conditions in the phototrophic bacteria is unknown. Presumably it serves as a carbon rather than an energy reserve and in accord with this, Stanier et al. (1959) showed that PHB could be converted into other cell constituents in the light in the presence of CO_2.

REFERENCES

ALBERS H. & GOTTSCHALK G. (1976) Acetate metabolism in *Rhodopseudomonas gelatinosa* and several other Rhodospirillaceae. *Arch. Microbiol.*, **111**, 45–49.

ANDERSON L. & FULLER R. C. (1967a) Photosynthesis in *Rhodospirillum rubrum* I. Autotrophic carbon dioxide fixation. *Plant Physiol.*, **42**, 487–90.

ANDERSON L. & FULLER R. C. (1967b) Photosynthesis in *Rhodospirillum rubrum*. II. Photoheterotrophic carbon dioxide fixation. *Plant Physiol.*, **42**, 491–6.

ANDERSON L. & FULLER R. C. (1967c) Photosynthesis in *Rhodospirillum rubrum* III. Metabolic control of reductive pentose phosphate and tricarboxylic cycle enzymes. *Plant Physiol.*, **42**, 497–502.

ANDERSON L., PRICE C. B. & FULLER R. C. Molecular diversity of the ribulose-1,5-diphosphate carboxylase from photosynthetic microorganisms. *Science*, **161**, 482–4.

ATKINSON D. E. (1968) The energy charge of the adenylate pool as a regulatory parameter. Interaction with feedback modifiers. *Biochemistry, N.Y.*, **7**, 4030–4.

BACHOFEN R., BUCHANAN B. B. & ARNON D. I. (1964) Ferredoxin as a reductant in pyruvate synthesis by a bacterial extract. *Proc. natn. Acad. Sci. USA*, **51**, 690–4.

BADGER M. R. (1980) Kinetic properties of ribulose 1,5-bisphosphate carboxylase/oxygenase from *Anabaena variabilis*. *Arch. Biochem. Biophys.*, **201**, 247–54.

BEATTY J. T. & GEST H. (1981) Generation of succinyl-coenzyme A in photosynthetic bacteria. *Arch. Microbiol.*, **129**, 335–40.

BEUSCHER N. & GOTTSCHALK G. (1972) Lack of citrate lyase—the key enzyme of the reductive carboxylic acid cycle—in *Chlorobium thiosulfatophilum* and *Rhodospirillum rubrum*. *Z. Naturforsch.*, **27b**, 967–73.

BOATMAN E. (1964) Observations on the fine structure of sphaeroplasts of *Rhodospirillum rubrum*. *J. Cell Biol.*, **20**, 297–311.

BUCHANAN B. B., EVANS M. C. W. & ARNON D. I. (1967) Ferredoxin-dependent carbon assimilation in *Rhodospirillum rubrum*. *Arch. Microbiol.*, **59**, 32–40.

BUCHANAN B. B., SCHURMANN P. & SHANMUGAM K. T. (1972) Role of the reductive carboxylic acid cycle in a photosynthetic bacterium lacking ribulose 1,5-diphosphate carboxylase. *Biochim. biophys. Acta*, **283**, 136–45.

BUCHANAN B. B. & SIREVÅG R. (1976) Ribulose 1,5-diphosphate carboxylase and *Chlorobium thiosulfatophilum*. *Arch. Microbiol.*, **109**, 15–19.

CANDY D. S. (1980) *Biological Function of Carbohydrates.* pp. 197. Blackie, Glasgow & London.

COHEN-BAZIRE G. & KUNISAWA R. (1963) The fine structure of *Rhodospirillum rubrum*. *J. Cell Biol.*, **16**, 401–19.

CUTINELLI C., EHRENSVÄRD G., REIO L., SALUSTE E. & STJERNHOLM R. (1951) Acetic acid metabolism in *Rhodospirillum rubrum* under anaerobic condition. II. *Ark. Kemi*, **3**, 315–22.

DUTTON P. L. & EVANS W. C. (1969) The metabolism of aromatic compounds by *Rhodopseudomonas palustris*: a new reductive method of aromatic ring metabolism. *Biochem. J.*, **113**, 525–36.

ELSDEN S. R. & ORMEROD J. G. (1956) The effect of monofluoroacetate on the metabolism of *Rhodospirillum rubrum*. *Biochem. J.*, **63**, 691–701.

EVANS M. C. W. (1965) The photoassimilation of succinate to hexose by *Rhodospirillum rubrum*. *Biochem. J.*, **95**, 669–77.

EVANS M. C. W. & BUCHANAN B. B. (1965) Photoreduction of ferredoxin and its use in carbon dioxide fixation by a subcellular system from a photosynthetic bacterium. *Proc. natn. Acad. Sci. USA*, **53**, 1420–5.

EVANS M. C. W., BUCHANAN B. B. & ARNON D. I. (1966) A new ferredoxin dependent carbon reduction cycle in a photosynthetic bacterium. *Proc. natn. Acad. Sci. USA*, **55**, 928–34.

FUCHS G., STUPPERICH E. & JAENCHEN R. (1980a) Autotrophic CO_2 fixation in *Chlorobium limicola*. Evidence against the operation of the Calvin cycle in growing cells. *Arch. Microbiol.*, **128**, 56–63.

FUCHS G., STUPPERICH E. & EDEN G. (1980b) Autotrophic CO_2 fixation in *Chlorobium limicola*. Evidence for the operation of a reductive tricarboxylic acid cycle in growing cells. *Arch. Microbiol.*, **128**, 64–71.

FULLER R. C. (1978) Photosynthetic carbon metabolism in the green and purple bacteria. In *The Photosynthetic Bacteria* (Ed. by R. K. Clayton & W. R. Sistrom) pp. 691–705. Plenum Press, New York and London.

FULLER R. C., SMILLIE R. H., SISLER E. C. & KORNBERG H. L. (1961) Carbon metabolism in *Chromatium*. *J. Biol. Chem.*, **236**, 2140–9.

GEMERDEN H. VAN (1968a) Utilization of reducing power in growing cultures of *Chromatium*. *Arch. Mikrobiol.*, **64**, 111–17.

GEMERDEN H. VAN (1968b) On the ATP generation by *Chromatium* in darkness. *Arch. Mikrobiol.*, **64**, 118–24.

GEST H. & KAMEN M. D. (1949) Photoproduction of molecular hydrogen by *Rhodospirillum rubrum*. *Science*, **109**, 558–9.

GEST H., ORMEROD J. G. & ORMEROD K. S. (1962) Photometabolism of *Rhodospirillum rubrum*: light-dependent dissimilation of organic compounds to carbon dioxide and molecular hydrogen by an anaerobic citric acid cycle. *Arch. Biochem. Biophys.*, **97**, 21–33.

GIBSON J. L. & TABITA R. F. (1977) Different molecular forms of D-ribulose-1,5-bisphosphate carboxylase from *Rhodopseudomonas sphaeroides*. *J. Biol. Chem.*, **252**, 943–9.

GLOVER J., KAMEN M. D. & VAN GENDEREN H. (1952) Studies on the metabolism of photosynthetic bacteria. XII. Comparative light and dark metabolism of acetate and carbonate by *Rhodospirillum rubrum*. *Arch. Biochem. Biophys.*, **35**, 384–408.

HOARE D. S. (1963) The photo-assimilation of acetate by *Rhodospirillum rubrum*. *Biochem. J.*, **87**, 284–301.

HOARE D. S. & GIBSON J. (1964) Photoassimilation of acetate and the biosynthesis of amino acids by *Chlorobium thiosulfatophilum*. *Biochem. J.*, **91**, 546–59.

IVANOVSKY R. N., SINSTOV N. V. & KONDRATIEVA E. N. (1980) ATP-linked citrate lyase activity in the green sulfur bacterium *Chlorobium limicola* forma *thiosulfatophilum*. *Arch. Microbiol.*, **128**, 239–41.

KNIGHT M. (1962) The photometabolism of propionate by *Rhodospirillum rubrum*. *Biochem. J.*, **84**, 170–85.

KOHLMILLER E. F. & GEST H. (1951) A comparative study of the light and dark fermentations of organic acids by *Rhodospirillum rubrum*. *J. Bact.*, **61**, 269–82.

LARSEN H. (1951) Photosynthesis of succinic acid by *Chlorobium thiosulfatophilum*. *J. Biol. Chem.*, **193**, 167–73.

LUNDGREN D. G., ALPER R., SCHNAITMAN C. & MARCHESSAULT R. H. (1965) Characterization of poly-β-hydroxybutyrate extracted from different bacteria. *J. Bact.*, **89**, 245–51.

MADIGAN M. T. & BROCK T. D. (1977) CO_2 fixation in photosynthetically-grown *Chloroflexus aurantiacus*. *FEMS Microbiol. Lett.*, **1**, 301–4.

MADIGAN M. T., PETERSEN S. R. & BROCK T. D. (1974) Nutritional studies on *Chloroflexus*, a filamentous photosynthetic gliding bacterium. *Arch. Microbiol.*, **100**, 97–103.

MCFADDEN B. A. (1978) Assimilation of one-carbon compounds. In *The Bacteria* Vol. VI. *Bacterial Diversity* (Ed. by I. C. Gunsalus), pp. 219–304. Academic Press, New York, San Francisco, London.

MOSKOWITZ G. J. & MERRICK J. M. (1969) Metabolism of poly-β-hydroxybutyrate. II. Enzymatic synthesis of D(-)-β-hydroxybutyryl coenzyme A by an enoyl hydrase from *Rhodospirillum rubrum*. *Biochemistry, N.Y.*, **8**, 2748–55.

ORMEROD J. G. (1956) The use of radioactive carbon dioxide in the measurement of carbon dioxide fixation in *Rhodospirillum rubrum*. *Biochem. J.*, **64**, 373–80.

ORMEROD J. G. (1980) Photosynthesis of acetate from carbon dioxide. Third International Symposium on Microbial Growth on C_1 Compounds. pp. 99–100. Sheffield, UK.

PAALME T., OLIVSON A. & VILU R. (1982) ^{13}C-NMR study of the glucose synthesis pathways in the bacterium *Chlorobium thiosulfatophilum*. *Biochim. Biophys. Acta*, **720**, 303–10.

QUANDT L. (1978) Thesis, University of Göttingen.

QUANDT L., GOTTSCHALK G., ZIEGLER H. & STICHLER W. (1977) Isotope discrimination by photosynthetic bacteria. *FEMS Microbiol. Lett.*, **1**, 125–8.

QUANDT L., PFENNIG N. & GOTTSCHALK G. (1978) Evidence for the key position of pyruvate synthase in the assimilation of CO_2 by *Chlorobium*. *FEMS Microbiol. Lett.*, **3**, 227–30.

QUAYLE J. R. & FERENCI T. (1978) Evolutionary aspects of autotrophy. *Microbiol. Rev.*, **42**, 251–73.

SADLER W. R. & STANIER R. Y. (1960) The function of acetate in photosynthesis by green bacteria. *Proc. natn. Acad. Sci. USA*, **46**, 1328–34.

SHIGESADA K., HIDAKA K., KATSUKI H. & TANAKA S. (1966) Biosynthesis of glutamate in photosynthetic bacteria. *Biochim. Biophys. Acta*, **112**, 182–5.

SIREVÅG R. (1974) Further studies on carbon dioxide fixation in *Chlorobium*. *Arch. Microbiol.*, **93**, 3–18.

SIREVÅG R. (1975) Photoassimilation of acetate and metabolism of carbohydrate in *Chlorobium thiosulfatophilum*. *Arch. Microbiol.*, **104**, 105–111.

SIREVÅG R., BUCHANAN B. B., BERRY J. A. & TROUGHTON J. H. (1977) Mechanisms of CO_2 fixation in bacterial photosynthesis studied by the carbon isotope fractionation technique. *Arch. Microbiol.*, **112**, 35–38.

SIREVÅG R. & CASTENHOLZ R. W. (1979) Aspects of carbon metabolism in *Chloroflexus*. *Arch. Microbiol.*, **120**, 151–3.
SIREVÅG R. & ORMEROD J. G. (1970a) Carbon dioxide fixation in green sulphur bacteria. *Biochem. J.*, **120**, 399–408.
SIREVÅG R. & ORMEROD J. G. (1970b) Carbon dioxide fixation in photosynthetic green sulphur bacteria. *Science*, **169**, 186–8.
SIREVÅG R. & ORMEROD J. G. (1977) Synthesis, storage and degradation of polyglucose in *Chlorobium thiosulfatophilum*. *Arch. Microbiol.*, **111**, 239–44.
SMILLIE R. M., RIGOPOULOS N. & KELLY H. (1962) Enzymes of the reductive pentose phosphate cycle in the purple and the green photosynthetic sulphur bacteria. *Biochim. Biophys. Acta*, **56**, 612–14.
STANIER R. Y., DOUDOROFF M., KUNISAWA R. & CONTOPOULOU R. (1959) The role of organic substrates in bacterial photosynthesis. *Proc. natn. Acad. Sci. USA*, **45**, 1246–60.
STOPPANI A. O. H., FULLER R. C. & CALVIN M. (1952) Carbon dioxide fixation by *Rhodopseudomonas capsulatus*. *J. Bact.*, **69**, 491–501.
TABITA R. F. (1981) Molecular regulation of carbon dioxide assimilation in autotrophic microorganisms. In *Microbial Growth on C_1 Compounds* (Ed. by H. Dalton) pp. 70–82. Heyden, London.
TABITA R. F., MCFADDEN B. & PFENNIG N. (1974) D-ribulose-1,5-bisphosphate carboxylase in *Chlorobium thiosulfatophilum*. *Biochim. Biophys. Acta*, **341**, 187–94.
TAKABE T. & AKAZAWA T. (1977) A comparative study on the effect of O_2 on photosynthetic carbon metabolism by *Chlorobium thiosulfatophilum* and *Chromatium vinosum*. *Plant and Cell Physiol.*, **18**, 753–65.
THAUER R. K., RUPPRECHT E. & JUNGERMANN K. (1970). Glyoxylate inhibition of clostridial pyruvate synthase. *FEBS Lett.*, **9**, 271–3.
WALKER D. A. & ROBERTSON S. P. (1978) Regulation of photosynthetic carbon assimilation. In *Photosynthetic Carbon Assimilation* (Ed. by H. W. Siegelman & G. Hind) pp. 43–59. Basic Life Sciences Vol. II. Plenum Press, New York and London.
WHELAN T., SACKETT W. M. & BENEDICT C. R. (1973) Enzymatic fractionation of carbon isotopes by phosphoenolpyruvate carboxylase from C_4 plants. *Plant Physiol.*, **51**, 1051–4.
WHITTLE P. J., LUNT D. O. & EVANS W. C. (1976) Anaerobic photometabolism of aromatic compounds by *Rhodopseudomonas* sp. *Biochem. Soc. Trans.*, **4**, 490–1.
WOLOSIUK R. A. & BUCHANAN B. B. (1977) Thioredoxin and glutathione regulate photosynthesis in chloroplasts. *Nature, Lond.*, **266**, 565–7.

Chapter 7. Nitrogen Fixation and Ammonia Assimilation

B. C. JOHANSSON, S. NORDLUND AND H. BALTSCHEFFSKY

7.1 INTRODUCTION

In nature the ammonium ion is probably the most common nitrogen source for phototrophic bacteria. Although this ion is also the best nitrogen source in the sense that it gives the highest growth rates, the widespread ability to fix dinitrogen amongst phototrophic bacteria strongly suggests that they are periodically, at least, exposed to conditions where the ammonium concentration limits growth. The utilization of more oxidized nitrogen sources such as nitrate by phototrophic bacteria seems to be exceptional and confined to some purple non-sulphur bacteria.

In nitrogen fixation dinitrogen, N_2, is reduced to ammonia in a complex reaction catalysed by the enzyme nitrogenase. The newly fixed nitrogen, i.e. the ammonium ion, is assimilated via the enzyme couple glutamine synthetase/glutamate synthase in all phototrophic bacteria examined so far (Fig. 7.1). A complex regulatory system for glutamine synthetase has been observed in heterotrophic bacteria and has been demonstrated also in some species of phototrophic bacteria.

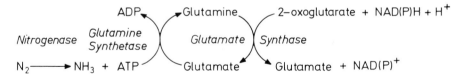

Fig. 7.1 The path of nitrogen during nitrogen fixation and ammonia assimilation in phototrophic bacteria.

The central role of glutamate as a precursor of other amino acids and as amino donor in transferase reactions is well known (Fig. 7.2). Likewise the amide group of glutamine plays a vital part in the synthesis of nucleotides and amino acids. Thus glutamate and glutamine are required to make the two most important high molecular weight components of living cells, the nucleic acids and the proteins.

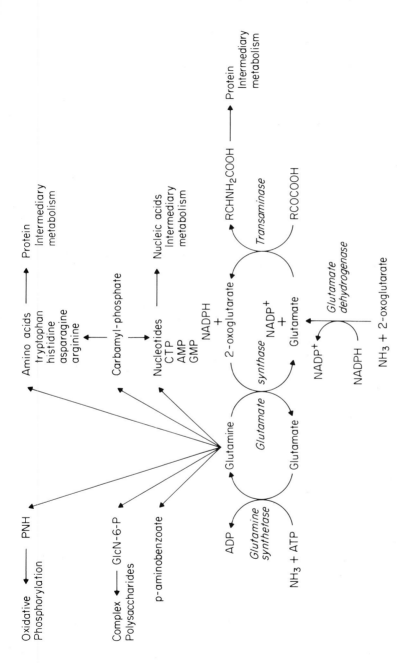

Fig. 7.2 Pathways of ammonia assimilation for the production of glutamate and glutamine in bacteria, and some of the roles of these compounds in intermediary metabolism. (From: Tyler, 1978. Reproduced, with permission, from the Annual Review of Biochemistry, Vol. 47. © 1978 by Annual Reviews Inc.)

Although the number of phototrophic bacteria shown to be able to fix nitrogen has increased over the years, the molecular properties of nitrogenase have been studied in detail in only three species. Nitrogenase from some species of Rhodospirillaceae shows, in addition to genetic regulation, some new regulatory features, which distinguish it from the enzyme isolated from other nitrogen fixing organisms. This metabolic regulation is the subject of intense investigation in several laboratories.

Genetic techniques are being used and will no doubt be used increasingly to study nitrogen fixation, ammonia assimilation and the possible regulatory connection between nitrogenase and glutamine synthetase in phototrophic bacteria. Recent results obtained with the recombinant DNA technique in studies of nitrogen fixation in non-phototrophic bacteria support this prediction.

This chapter starts with a description of nitrogen fixation, its enzyme nitrogenase, and its regulation. In a logical continuation, the assimilation of ammonia, the product from the nitrogen fixation reaction, is discussed, with special emphasis on the key enzyme glutamine synthetase. Finally the genetic studies of nitrogen fixation and ammonia assimilation in phototrophic bacteria are treated in some detail.

7.2 NITROGEN FIXATION

Nitrogen fixation in phototrophic bacteria was first demonstrated in 1949 in studies on hydrogen production by the purple non-sulphur bacterium *Rhodospirillum rubrum* (Kamen & Gest, 1949). This was soon followed by reports on nitrogen fixation in other phototrophic bacteria e.g. *Rhodopseudomonas sphaeroides, Rp. palustris, Rp. capsulata, Rp. gelatinosa, Chromatium vinosum, Rhodomicrobium vannielii* and *Chlorobium limicola* (Lindstrom et al., 1949; 1950; 1951; 1952). Siefert (1976) tested 52 strains, and found that among the three families studied, 24 of 26 Rhodospirillaceae, 16 of 19 Chromatiaceae and 13 of 17 Chlorobiaceae exhibited nitrogen fixing activity.

7.2.1 Some physiological aspects of nitrogen fixation

Early studies on photoproduction of hydrogen and nitrogen fixation showed that cells grown with N_2 or glutamate but not ammonia as nitrogen source were able to fix nitrogen (Gest et al., 1950). These authors also demonstrated that N_2-grown cells could produce substantial amounts of hydrogen, and that this hydrogen production could be inhibited by either ammonium ions or N_2.

A number of investigations on photoproduction of hydrogen and nitrogen fixation in whole cells, mainly *Rs. rubrum*, that were reported over the next ten years provided important information about the physiology of these processes. It was shown that cells grown in the presence of aspartate but not asparagine or glutamine would produce hydrogen (Bregoff & Kamen, 1952) and that nitrogen fixation was inhibited by low concentrations of oxygen (Pratt & Frenkel, 1959). It was also demonstrated that the synthesis of the enzyme(s) responsible for H_2-production became derepressed after the ammonia had been utilized in cultures of *Rs. rubrum* grown on limiting ammonia medium (Ormerod et al., 1961).

The close relationship between hydrogen production and nitrogen fixation demonstrated in a number of studies, led Ormerod & Gest (1962) to propose that both processes are catalysed by the same enzyme or enzyme complex. This was later shown to be the case in *Azotobacter vinelandii* by Bulen et al., (1965). These authors provided some evidence that this was also true for *Rs. rubrum*. Direct evidence showing that nitrogenase catalyses both processes in phototrophic bacteria was provided in genetic studies on *Rp. capsulata* by Wall et al. (1975). Since production of H_2 is an intrinsic activity of nitrogenase (see below), bacteria grown under nitrogen fixing conditions always have the capacity to evolve hydrogen, but exhibit this activity best in the absence of N_2 or in suspensions of resting cells. Production of H_2 by nitrogenase has been suggested to function as a 'safety valve' through which the cell can dispose of excess reducing power (Gest, 1972). Production of H_2 as well as nitrogen fixation by purple non-sulphur bacteria is rapidly inhibited by addition of ammonia to the culture. This phenomenon, which has been termed a 'switch-off' effect is further discussed in Section 7.2.3.

The early investigations of nitrogen fixation by purple bacteria indicated that it was strictly light-dependent and required anaerobic conditions. Over the last years, however, it has been shown that anaerobic sugar fermentation as well as aerobic respiration at low O_2-tensions support growth in the dark with N_2 as nitrogen source (Madigan et al., 1979; Siefert & Pfennig, 1980). In addition Ludden & Burris (1981) have demonstrated that *Rs. rubrum* reduces acetylene (the most commonly used assay for nitrogenase) at appreciable rates in the dark with pyruvate as substrate. Consequently there is no absolute requirement for light to produce ATP and to generate reductant needed for N_2-fixation. It should, however, be noted that the highest nitrogenase activities are obtained in cultures grown in the light with organic acids, e.g. malate, succinate or pyruvate, as electron donors. Purple and green sulphur bacteria also utilize inorganic compounds such as thiosulphate or H_2 as reductants (Ormerod et al., 1961; Klemme, 1968; Schick, 1971b).

7.2.2 Purification and properties of nitrogenase

In 1960 Carnahan and co-workers described a procedure for obtaining cell-free extracts with high nitrogenase activity from *Clostridium pasteurianum*

(Carnahan et al., 1960). This breakthrough was followed by reports of active cell-free extracts from *Rs. rubrum* (Schneider et al., 1960) and *Chromatium* (Arnon et al., 1960).

The nitrogenases of phototrophic bacteria, like all other nitrogenases, consist of two proteins, the MoFe-protein (dinitrogenase) and the Fe-protein (dinitrogenase reductase). These have been extensively purified from three phototrophic bacteria: *Chromatium*, *Rs. rubrum* and *Rp. capsulata*. Evans and co-workers purified the MoFe-protein from *Chromatium* to near homogeneity and obtained a partial purification of the Fe-protein (Evans et al., 1973). The MoFe-protein was also partially characterized. In addition the MoFe-protein and the Fe-protein from '*Chloropseudomonas ethylica*' were separated and used in cross-reaction experiments (Smith et al., 1971). Cultures designated '*Chl. ethylica*' are now known to be mixtures of green phototrophic bacteria and sulphur- or sulphate-reducing bacteria (Olson, 1978) and it has not been rigorously demonstrated that the nitrogenase isolated originated from the phototrophic bacterium.

Both the MoFe-protein and the Fe-protein from *Rs. rubrum* have been purified to near homogeneity and characterized in some detail (Nordlund et al., 1978; Ludden & Burris, 1978). The purification procedures involve traditional separation techniques. All steps must, however, be made under anaerobic conditions since both of the nitrogenase proteins are oxygen sensitive. Some molecular properties of nitrogenase from phototrophic bacteria are shown in Table 7.1. The properties of the enzyme from *Clostridium pasteurianum* are included for comparison. It is evident that nitrogenase of the phototrophic bacteria has properties very similar to that of non-phototrophic bacteria. In support of this are results showing that the separated *Rs. rubrum* proteins cross-react with nitrogenase components from a great number of other bacteria (Biggins et al., 1971; Emerich & Burris, 1978).

Nitrogenase from phototrophic bacteria like that from other bacteria requires MgATP and an electron donor to catalyse the reduction of substrates. ATP is hydrolysed to ADP and phosphate and the minimum number of ATP hydrolysed per $2e^-$ transferred by *Rs. rubrum* nitrogenase *in vitro* is 4 (Nordlund & Eriksson, 1979) which is the same as the value for non-phototrophic bacteria (Mortenson & Thorneley, 1979).

The physiological electron donor for nitrogenase of phototrophic bacteria has not yet been identified but it has been demonstrated that ferredoxin I and II from *Rs. rubrum* can transfer reducing equivalents from illuminated chloroplasts to a crude preparation of *Rs. rubrum* nitrogenase (Yoch & Arnon, 1975). It was also shown that ferredoxin I was more effective than ferredoxin II. Recently it has been demonstrated that pyruvate, 2-oxoglutarate and, to a lesser extent, oxaloacetate support nitrogenase activity in crude extracts of *Rs. rubrum* (Ludden & Burris, 1981). It was also shown that added *Rs. rubrum* ferredoxin greatly enhanced the pyruvate-driven activity, supporting the proposal that ferredoxin is the physiological electron donor to nitrogenase

Table 7.1 Comparison of nitrogenase from *Rs. rubrum*, *Rp. capsulata*, *Chromatium* and *Clostridium pasteurianum*.

	Mol. wt.	Subunits	Mo	Fe	S	Spec. act nmol/min.mg	Ref
Rs. rubrum							
MoFe-protein	215 000	4 × 56 000	2,	25–30	19–22	920	a
	230 000	4 × 58 500	1,7	20		1725	b
Fe-protein	65 000	2 × 31 500		3.8–4.5	2.8–5.0	1260	a, c
	61 500	30 000 + 31 500		3.5		845	b
Rp. capsulata							
MoFe-protein	230 000	2 × 59 500	1,2–1,4	23.4–32	24,5–77,7	1800	d
		2 × 55 000					
Fe-protein	63 000	2 × 33 500	—	3,2–5	4,5–5,5	850	d
Chromatium							
MoFe-protein			1.3*	18.7*	15.2*	1800	e
Cl. pasteurianum							
Mo-Fe-protein	220 000	2 × 50 000	2	30 + 2	24–30	2500	f, g
		2 × 60 000					
Fe-protein	57 000	2 identical		4	4	3100	f, g

* the metal content has been recalculated on the assumption of the molecular weight 220 000.
(a) Nordlund *et al.*, 1978; (b) Ludden & Burris, 1978; (c) Nordlund & Eriksson, 1979; (d) Hallenbeck *et al.*, 1982; (e) Evans *et al.*, 1973; (f) Mortenson & Thorneley, 1979; (g) Tso, 1974.

although it is not properly understood how reduced ferredoxin is generated in purple bacteria growing in the light (see Chapter 4). The other plausible electron donor, flavodoxin, has only been isolated from cells grown in iron-deficient medium (Cusanovich & Edmondson, 1971). The most common electron donor used *in vitro* is sodium dithionite.

In summary, the overall reaction catalysed by nitrogenase can be written as:

$$N_2 + 12\,MgATP + 6\,(H^+ + e^-) \rightarrow 2\,NH_3 + 12\,(MgADP + P_i).$$

In addition to the reduction of dinitrogen to ammonia, nitrogenase catalyses a number of other reductions. Munson & Burris (1969) demonstrated that a crude preparation of *Rs. rubrum* nitrogenase catalysed the reduction of acetylene, protons, cyanide, isocyanide and azide, all of which have been shown to be substrates of nitrogenase from non-phototrophic bacteria. Of these reactions, only the production of hydrogen is of physiological importance. Even in the presence of nitrogen some electrons are diverted to the production of hydrogen *in vivo*. *In vitro* studies using *Rs. rubrum* nitrogenase showed that the amount of hydrogen produced in the presence of acetylene by purified nitrogenase components was dependent on the ratio between the MoFe-protein and the Fe-protein. Increasing this ratio resulted in increased hydrogen production (Nordlund & Eriksson, 1979). This phenomenon, which has also been documented in non-phototrophic bacteria, may be due to the MoFe-protein being less reduced when it is present in excess or when the ATP concentration is low (Nordlund & Eriksson, 1979; Hageman & Burris, 1980).

Carbon monoxide has long been known to inhibit nitrogen fixation (Lind & Wilson, 1941) and it has been shown to inhibit all reductions catalysed by nitrogenase except hydrogen production (Hardy & Burns, 1968). In keeping with this, the addition of CO to *Rs. rubrum* nitrogenase inhibited acetylene reduction and increased the concomitant production of hydrogen (Nordlund & Eriksson, 1979).

7.2.3 Activation of the Fe-protein and metabolic regulation of nitrogen fixation

As already indicated, the synthesis of nitrogenase in *Rs. rubrum* is repressed by ammonia in the growth medium. The mechanism of this genetic regulation is not yet known in detail in phototrophic bacteria.

The nitrogenase of some phototrophic bacteria is known to be regulated also at the metabolic level, as shown by the fact that addition of ammonium ions to a nitrogen fixing culture leads to rapid inhibition of nitrogenase activity. This effect, first demonstrated in *Rs. rubrum* by Gest et al. (1950), has been documented more recently in several reports (Schick, 1971a; Neilson & Nordlund, 1975; Sweet & Burris, 1981) and termed the 'switch-off' effect

(Zumft & Castillo, 1978). Nitrogenase activity is regained when the ammonia has been metabolized.

The 'switch-off' effect, which has only been produced *in vivo*, has also been shown in *Rp. capsulata* (Hillmer & Gest, 1977; Meyer *et al.*, 1978a), *Rp. palustris* (Zumft & Castillo, 1978; Kim *et al.*, 1980), *Rp. sphaeroides* (Jones & Monty, 1979), *Rp. acidophila* (Siefert, 1976) and *Rhodomicrobium vannielii* (Zumft *et al.*, 1981) and may be a general feature in nitrogen-fixing *Rhodospirillaceae*. The metabolic regulation of nitrogenase was also indicated by the early investigations on the cell-free nitrogenase system from *Rs. rubrum*. Thus there were difficulties in obtaining extracts with high activity (Burns & Bulen, 1966; Munson & Burris, 1969), and it was found that an unusually high Mg^{2+}:ATP ratio (25:5) was required to obtain maximal activity *in vitro* and that chromatophores were inhibitory (Burns & Bulen, 1966). Munson & Burris (1969) in addition reported that the time-course of the nitrogenase reaction exhibited a lag which could be abolished by passing the extract through a Sephadex G-25 column. These observations are now known to be due indeed to the existence of an interesting regulatory system in *Rs. rubrum*. The investigation of this system is now described and evidence presented that it constitutes an important part of the reversible 'switch-off' effect.

When nitrogenase is isolated from *Rs. rubrum* cultures grown with N_2 or glutamate as nitrogen source, it exhibits low or no activity, although nitrogenase activity *in vivo* is high. The low activity of the extracts is believed to be due to the Fe-protein being converted to the inactive form during harvesting and/or breaking of the cells. The fully active Fe-protein can be isolated from cells grown under nitrogen-starvation. If, however, ammonia is added to such cultures prior to harvesting, the Fe-protein is converted to the inactive form (Carithers *et al.*, 1979; Nordlund & Eriksson, 1979).

The inactive Fe-protein can be activated *in vitro* by an activating enzyme, which can be solubilized from chromatophores with 0.5 M NaCl (Baltscheffsky & Nordlund, 1975). This enzyme, which is very sensitive to oxygen, has been partially purified under anaerobic conditions (Zumft & Nordlund, 1981). The inclusion of 0.5 mM Mn^{2+} in the buffers used in the purification greatly stabilized the enzyme. The molecular weight of the activating enzyme is 20 000.

The activating enzyme is present also in cells grown with ammonia as nitrogen source (Ludden & Burris, 1976; Nordlund *et al.*, 1977) and even in aerobically grown cells (Nordlund, unpublished). These observations raise the question of whether this enzyme has other functions in the cell.

The activation, which requires ATP, is also dependent on Mn^{2+} (Ludden & Burris, 1976; Nordlund *et al.*, 1977). Yoch (1979) has suggested that *in vivo* the activating enzyme is dependent on Mn^{2+}.

The molecular events taking place on the Fe-protein during activation have been clarified in some detail by the work of Ludden & Burris (1978; 1979) (Fig. 7.3). They have shown that the inactive form of the Fe-protein is covalently modified by a group containing pentose, phosphate and an 'adenine-like'

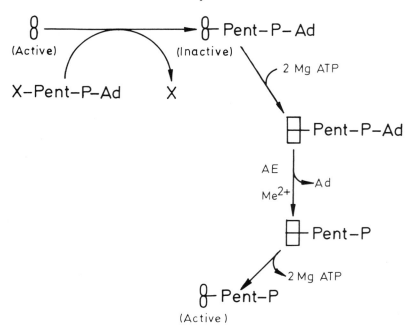

Fig. 7.3 Hypothesis for the mechanism of inactivation and activation of the *Rs. rubrum* Fe-protein. The inactivation reaction is initiated by addition of NH_4^+-ions *in vivo* or during harvesting of cells grown with glutamate or N_2 as nitrogen source. Pent, pentose; P, Phosphate; Ad, adenine-like moiety; X, donor of modifying group; 0, □, subunit of the Fe-protein; AE, activating enzyme (modified version of scheme in Ludden & Burris, 1979).

molecule in the ratio 1:1:1 (Ludden, personal communication). Upon activation of the Fe-protein by the activating enzyme the 'adenine-like' molecule is removed leaving the pentose and the phosphate. No other enzyme, e.g. alkaline phosphatase or phosphodiesterase, has been shown to catalyse the activation of the inactive Fe-protein. In a recent report it was shown that the inactive Fe-protein from *Rp. capsulata* is also modified in the same manner (Hallenbeck *et al.*, 1982).

There are two other known differences between the modified and the unmodified Fe-protein apart from the activity: first, the UV-absorption spectrum of the inactive form has a shoulder at 268 nm, absent in the active form; secondly, SDS-polyacrylamide-gel electrophoresis of the inactive form produces two bands differing in molecular weight by about 1500 daltons. The intensity of staining of the two bands is about 1:1 changing to 1:4 upon activation, with the lower band dominating (Ludden & Preston, 1981). It was also reported that when the inactive form of the Fe-protein, isolated from cells grown in the presence of ^{32}P, was run on an SDS-gel and autoradiographed, only the upper band was labelled.

The activating enzyme has as yet only been demonstrated in *Rs. rubrum* and *Azospirillum lipoferum*, a non-phototrophic bacterium (Ludden et al., 1978). However, indirect evidence for such an enzyme in *Rp. capsulata* (Yoch, 1979; Zumft et al., 1981), *Rp. palustris* and *Rhodomicrobium vannielii* (Zumft et al., 1981) has been presented.

Thus far, the evidence from work with cell-free nitrogenase of *Chromatium* has given no indication of this type of regulation in this organism (Yoch & Arnon, 1970; Winter & Arnon, 1970; Winter & Ober, 1973; Evans et al., 1973).

Evidence that the regulatory system described above for *Rs. rubrum* is closely connected with the reversible 'switch-off' effect has been presented by Nordlund & Eriksson (1979) and Carithers et al., (1979). In this work nitrogenase isolated from cells to which ammonium ions had been added was shown to be inactive, and addition of activating enzyme, ATP and Mn^{2+} caused reappearance of activity. This indicates, then, on a molecular level, that the 'switch-off' effect is due to conversion of the active form of the Fe-protein to the inactive.

An alternative, more elaborate model than the above has, however, been proposed. It involves three different forms of the nitrogenase complex with different physical properties (Carithers et al., 1979; Yoch & Cantu, 1980). However, in the experiments on which this model is based, only crude or partially purified preparations of nitrogenase and the activating enzyme were used. In addition, results from *in vivo* studies on the 'switch-off' effect (Sweet & Burris, 1981) as well as investigations on complex formation between the MoFe-protein and the Fe-protein (active as well as inactive; Nordlund & Ludden, unpublished) are not in agreement with this model.

The sequence of molecular events leading to inactivation of the Fe-protein by the 'switch-off' effect is at present not understood, but may involve glutamine synthetase (Hillmer & Fahlbush, 1979; Yoch, 1980; Yoch & Cantu, 1980). This involvement may be indirect, through the production of glutamine from ammonia, as suggested by the fact that methionine sulphoximine, an inhibitor of glutamine synthetase, prevents 'switch-off' caused by ammonia or asparagine, but not that caused by glutamine (Nordlund & Eklund, 1979; Jones & Monty, 1979; Sweet & Burris, 1981; see also Section 7.4).

It has also been suggested that the 'switch-off' effect in *Rp. sphaeroides* is due to a change in the $\Delta\Psi$ component of the proton motive force (Veeger et al., 1981) which, it is presumed, affects electron transfer to nitrogenase and thus decreases the activity. This model does not explain the finding that not only ammonium ions but also glutamine produces inhibition of nitrogenase. This is important since these two compounds have opposite charge at physiological pH. In addition, this model does not explain the inactivation of the Fe-protein.

It is clear that the more complete explanation of the 'switch-off' effect will have to await the identification of the moiety rendering the Fe-protein inactive as well as the isolation of the component(s) of the system that is responsible for this regulation of nitrogenase on a metabolic level.

7.3 AMMONIA ASSIMILATION

7.3.1 Pathways of ammonia assimilation

Among the sources of inorganic nitrogen that can be used by phototrophic bacteria, ammonia supports the most rapid growth. Growth on molecular nitrogen is generally much slower. In bacteria there are essentially two different pathways for assimilating ammonia supplied from the environment or generated intracellularly (Tyler, 1978). Depending on the concentration of ammonia available either of the two pathways might be used. At low concentrations of ammonia (< 1 mM) the glutamine synthetase (E.C.6.3.1.2.) glutamate synthase (E.C.1.4.13.) pathway for the synthesis of glutamate is of central importance (Figs. 7.1 & 7.2). This pathway is ATP-dependent and physiologically irreversible. The low K_m for ammonia of glutamine synthetase (Table 7.2) makes possible the assimilation of ammonia at the low concentrations available during nitrogen fixation.

Table 7.2 K_m (NH_4^+) values for glutamine synthetase (GS), glutamate dehydrogenase (GDH) and alanine dehydrogenase (ADH) for some phototrophic bacteria.

Species	GS (mM)	GDH (mM)	ADH (mM)	Ref
Rs. rubrum	0.55	11.5		a
Rp. capsulata	< 1		8.3	b
Rp. acidophila	0.38			c
Chromatium vinosum (D)	0.32	14		d
Rp. palustris	0.4 (GS-AMP) 0.2 / 1.5 (GS)			e
Chlorobium limicola	0.55	13.5		e

(a) Brown & Herbert, 1977b; (b) Johansson & Gest, 1976; (c) Herbert *et al.*, 1978; (d) Brown & Herbert, 1977a; (e) Alef & Zumft, 1981.

The electron donor used in the glutamate synthase reaction may vary in the different species of phototrophic bacteria. All the purple and green sulphur bacteria examined so far use NADH as electron donor (Brown & Herbert, 1977a). Among the purple non-sulphur bacteria some use NADPH and others NADH (Table 7.3) (Brown & Herbert, 1977b).

When the concentration of ammonium salts is above 1 mM, some phototrophic bacteria contain glutamate dehydrogenase. This enzyme has a high K_m for ammonia (Table 7.2) and might be used together with, or alternatively to, the glutamine synthetase/glutamate synthase reaction for ammonia assimilation. As with the glutamate synthase reaction, the electron donor for glutamate dehydrogenase is either NADPH or NADH, depending on the bacterial species (Table 7.4).

Table 7.3 Nucleotide specificity of glutamate synthase from some species of *Rhodospirillaceae* (*Rs.*).

Species	Nucleotide	Ref.
Rs. rubrum	NADPH	a
Rs. fulvum	NADH	a
Rs. molischianum	NADH	a
Rhodomicrobium vannielii	NADH	a
Rp. sphaeroides	NADPH	a
Rp. capsulata	NADPH	b
Rp. palustris	NADPH, NADH	a
Rp. acidophila	NADH	c

(a) Brown & Herbert, 1977b; (b) Johansson & Gest, 1976; (c) Herbert et al., 1978.

Table 7.4 Nucleotide specificity of glutamate dehydrogenase.

Species	Nucleotide	Ref.
Rs. rubrum	NADH	a
Rhodomicrobium vannielii	NADPH	a
Rp. sphaeroides	NADPH (NADH)	a
Chromatium vinosum, strain D	NADH	b
Chlorobium limicola	NADH	b

(a) Brown & Herbert, 1977b; (b) Brown & Herbert, 1977a.

Glutamate dehydrogenase activity is very low or undetectable in cultures growing on molecular nitrogen, but high when an excess of ammonia is present. On the other hand, glutamine synthetase activity is usually higher when the ammonia concentration is limiting for growth.

In some species, i.e. *Rp. capsulata*, *Rs. fulvum* and *Rp. acidophila* no glutamate dehydrogenase activity has been detected under any growth condition (Johansson & Gest, 1976; Brown & Herbert, 1977b; Herbert et al., 1978). It is noteworthy that certain cyanobacteria (Neilson & Doudoroff, 1973) and *Bacillus* (Hong et al., 1959) species also appear to lack glutamate dehydrogenase. Such organisms usually manifest alanine dehydrogenase activity, and it has been suggested that this could account for ammonia assimilation under certain conditions. Alanine dehydrogenase activity was indeed observed in crude cell-free extracts of both *Rp. capsulata* and *Rp. acidophila* (Johansson & Gest, 1976; Herbert et al., 1978). Probably the role of this enzyme in ammonia assimilation is rather limited in the phototrophic bacteria, its main function being the deamination of alanine, when this amino acid is used as the only source of carbon and nitrogen. In *Rp. capsulata*, alanine dehydrogenase could be important in ammonia assimilation when pyruvate is the carbon source and ammonia is present at high concentration. Formation of glutamate under such growth conditions could occur, at least in part by coupling of the aminating alanine dehydrogenase reaction to L-alanine-2-oxoglutarate aminotransferase activity, which is present in *Rp. capsulata* (Johansson & Gest, 1976). This pathway to glutamate is not possible in *Rp. acidophila*, however, since this species seems to lack L-alanine-2-oxoglutarate aminotransferase activity (Herbert *et al.*, 1978). The lack of this enzyme and of glutamate dehydrogenase in *Rp. acidophila* is consistent with the finding that this bacterium does not grow with glutamate as sole nitrogen source. It is probable that the glutamine synthetase/glutamate synthase pathway is normally of major importance for the photosynthetic assimilation of both molecular nitrogen and ammonia in *Rp. capsulata* and *Rp. acidophila*.

7.3.2 Purification and molecular properties of glutamine synthetase

The glutamine synthetase of phototrophic bacteria was only studied in crude extracts before partial purification of the enzyme from *Rp. capsulata* was

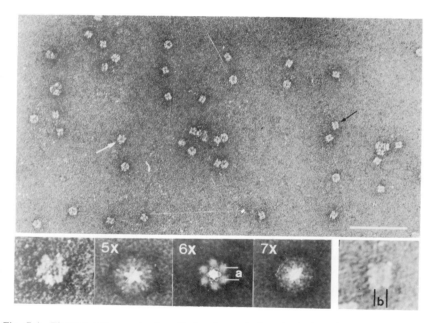

Fig. 7.4 Electron micrograph of purified and negatively stained glutamine synthetase from *Rhodopseudomonas palustris*. The white bar indicates 0.1 μm. The black arrow points to a side-on molecule. The white arrow indicates the face-on molecule which was analysed by the Markham rotation technique in the lower row (magnification 610 000 x). On six-fold rotation the hexameric symmetry became clearly visible; five-fold and seven-fold rotations served as controls. The subunit distance (a) within the plane of the hexameric subunit assembly was 5.4 nm; the distance between two planes (b) was 5.7 nm. (From: Alef *et al.*, 1981b).
Reproduced, with permission, from Zeitschrift für Naturforschung, Volume 36 c. © 1981 by Verlag der Zeitschrift für Naturforschung.

performed (Johansson & Gest, 1977). Glutamine synthetase in crude extracts of this bacterium can be completely sedimented by centrifugation at 140 000 × g for 2 hours. Further purification on a sucrose density gradient gave a preparation that was about 80-fold purified compared to the crude extract. It is noteworthy that about 50% of the glutamine synthetase of *Rp. sphaeroides* was in the pellet following a 100 000 × g centrifugation (1 h) (Lepo *et al.*, 1979). The enzyme from this organism was purified using an Affigel Blue column.

A very rapid and simple procedure has been worked out for the purification of glutamine synthetase from *Rs. rubrum* (Soliman *et al.*, 1981). Affinity chromatography on a column of ADP-agarose was used, after the cells had been treated with the detergent cetyltrimethylammonium bromide (CTAB). The CTAB-treatment is important, since membranes as well as proteins that interfere with the elution of the enzyme from the affinity column are removed after centrifugation of the broken cells. Using this procedure a 36-fold purification was obtained, with a recovery of 62%. The specific activities of the purified enzyme are comparable to the values for homogeneous glutamine synthetase from *Escherichia coli*. Preliminary results from molecular weight

Table 7.5 Molecular properties of glutamine synthetase from *Rhodopseudomonas palustris*.

Molecular weight from pore gradient PAGE*		Proposed subunit assembly
Highest value	582 000	α_{12}
Lowest value	57 900	α (monomer)
Intermediate values	123 000	α_2
	209 000	α_4
	288 000	α_6
Molecular weight from detergent PAGE	55 800	α (monomer)
No. of types of monomer	One	
Quarternary structure from electrophoretic analysis and electron microscopy	Dodecamer	
Molecular dimensions	(mean μ S.D.)	Sample size
Diameter (d)	15.3 μ 0.13 nm	29
Height (h)	11.1 μ 0.47 nm	12
Subunit distances† (a)	5.1 μ 0.08 nm	
(b)	5.5 μ 0.33 nm	

* PAGE = Polyacrylamide gel electrophoresis.
† Assuming an ideal hexametric structure these values were calculated as $a = d/3$, and $b = h/2$.

From: Alef *et al.*, 1981b. Reproduced, with permission, from Zeitschrift für Naturforschung, Volume 36 c. © 1981 by Verlag der Zeitschrift für Naturforschung.

determinations of the purified enzyme indicate a value between 550 000 and 580 000 daltons, which is similar to the values reported earlier for non-phototrophic bacteria (Stadtman & Ginsburg, 1974). The molecular weight of the subunit, as estimated by sodium dodecylsulphate gel electrophoresis, is around 50 000.

Glutamine synthetase from detergent treated *Rp. palustris* has been purified to homogeneity by using a column of Blue Sepharose Cl-6B (Alef *et al.*, 1981b) followed by chromatography on DEAE-cellulose. Electron microscopic examination of negatively stained enzyme preparations shows ring-like molecules, interspersed with more rectangular, bilayered structures (Fig. 7.4). The diameter of the molecule and the subunit distances are slightly greater than equivalent dimensions of the *E. coli* enzyme. Molecular properties of the enzyme from *Rp. palustris* are summarized in Table 7.5. Amino acid analysis showed that there is a considerable homology between the *Rp. palustris* enzyme and those from other bacteria, a cyanobacterium, a fungus and two higher plants (Alef *et al.*, 1981b).

7.3.3 Regulation of glutamine synthetase

Studies of glutamine synthetase from *E. coli* have shown that the enzymic activity is regulated by several different mechanisms:

1. repression and derepression of enzyme synthesis in response to variation in the concentration of ammonium salts;
2. interconversion of relaxed (inactive) and taut (active) forms, in response to variations in the concentration of divalent cations;
3. cumulative feedback inhibition by various end products of glutamine metabolism; and
4. covalent modification of the enzyme by the reversible adenylylation of a specific tyrosyl residue on each subunit (Stadtman *et al.*, 1970).

Of these control mechanisms, the cumulative feedback inhibition has been studied with the enzyme from *Rp. capsulata* (Johansson & Gest, 1976). The most notable inhibition was obtained with L-alanine, but D-alanine, carbamyl phosphate, glycine, L-histidine and L-tryptophan were also inhibitory. In contrast to most other glutamine synthetases that have been investigated, adenosine 5'-monophosphate had little or no effect on the *Rp. capsulata* enzyme. Feedback inhibition has also been observed with glutamine synthetase form *Rp. palustris* (Alef & Zumft, 1981). It was found that the deadenylylated form of the enzyme is preferentially sensitive to inhibition by L-alanine, glycine and L-serine. Adenosine 5'-monophosphate also had little effect on the *Rp. palustris* enzyme.

Adenylylation/deadenylylation is a major aspect of regulation of glutamine synthetase in a number of Gram-negative heterotrophic organisms (Tronick *et* al., 1973). The adenylylation state is dependent on the concentration of ammonia which in turn affects the concentration ratio between 2-oxoglutarate and glutamine. In *E. coli* this regulation is dependent on a complex cascade mechanism, consisting of at least two enzymes and a regulatory protein. Maximum biosynthetic activity is obtained when the enzyme is completely deadenylylated. However, the fully adenylylated, biosynthetically inactive enzyme, retains γ-glutamyl-transferase activity:

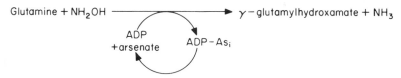

This transferase reaction is a non-physiological expression of glutamine synthetase activity and is frequently used in glutamine synthetase estimations. In the transferase assay, a convenient way of estimating the degree of adenylylation is to add Mg^{2+}, which inhibits only the adenylylated subunits. If activity is measured at the iso-activity pH for adenylylated and deadenylylated enzyme, the transferase assay is a measure of the total activity present in an extract, regardless of the degree of adenylylation. Another enzymic change occurring during adenylylation/deadenylylation is a shift in metal-specificity in the biosynthetic reaction. The deadenylylated enzyme is maximally stimulated by Mg^{2+}, but the adenylylated form requires Mn^{2+} for maximal activity.

These changes in enzymic activity can be used to detect the possible presence of adenylylation control. Deadenylylation of adenylylated enzyme can be accomplished *in vitro* by treating purified glutamine synthetase with snake venom phosphodiesterase. This hydrolyses the covalent linkage between AMP and the tyrosine residue in the glutamine synthetase molecule. During this cleavage any changes in enzymic activity can be recorded. This has been performed with the enzyme from *Rp. capsulata*. Significant changes in both the transferase and biosynthetic reactions were observed, which indicated the presence of adenylylation control in this organism (Johansson & Gest, 1977). Similar results have been obtained with *Rp. palustris* glutamine synthetase (Alef & Zumft, 1981), *Rp. sphaeroides* (permeabilized cells; Sakhno et al., 1981) and with partially purified enzyme from *Rs. rubrum* (Davies & Ormerod, 1982). In the last mentioned report glutamine synthetase of the green sulphur bacterium *Chlorobium limicola* was also tested, but no evidence for adenylylation was found.

Another test for this type of control is to compare the enzymic properties of glutamine synthetase from cells grown with excess and limiting ammonia as nitrogen source. When this was investigated with *Rp. capsulata*, no differences could be detected between glutamine synthetase activities measured in extracts from cells grown under these two conditions (Johansson & Gest, 1976).

Subsequently it was found that changes in the adenylylation state of the enzyme took place during harvesting of the cells. This could be prevented by adding CTAB to the cultures before harvesting. This treatment, which inhibits adenylyl transferase (Mura & Stadtman, 1981) resolved the different enzymic properties of glutamine synthetase, depending on the nitrogen source used for growth (Johansson & Gest, 1977). These differences are in accordance with those observed with glutamine synthetase from *E. coli*. Evidence for such metabolic interconversion of the enzyme from *Rp. palustris*, *Rp. sphaeroides* and *Rs. rubrum* has also recently been obtained (Alef & Zumft, 1981; Sakhno *et al.*, 1981; Davies & Ormerod, 1982).

The adenylylation state of glutamine synthetase is also affected by the light intensity during growth. Thus, in *Rp. capsulata* growing at low light intensity, the enzyme was largely adenylylated (Johansson & Gest, 1977), although the cells were growing in a medium (with glutamate as nitrogen source) that yields a deadenylylated enzyme at higher light intensities. On shifting from high to low light intensity (shift-down) the extent of adenylylation began to increase rapidly and the conversion was essentially complete in 25 minutes. Addition of FCCP (carbonyl-cyanide-p-trifluoromethoxyphenylhydrazone), a potent uncoupler of photophosphorylation, to cells growing at saturating light intensity mimics the effect of a shift-down in light intensity with respect to increasing the extent of adenylylation. Conversely, the effect of a shift-up in light intensity was to decrease the extent of adenylylation, but there is a lag of about 20 minutes before deadenylylation begins, and the conversion is not complete until about 60 minutes. Possibly the light intensity effects are due to regulation of adenylylation/deadenylylation by the 'energy state' of the cell. Interestingly, nitrogenase biosynthesis by *Rp. capsulata* also has been shown to be dependent on light (Meyer *et al.*, 1978b; see, however, Section 7.2.1).

So far *Rp. capsulata*, *Rp. palustris*, *Rp. sphaeroides* and *Rs. rubrum* are the only phototrophic bacteria in which there is evidence for regulation of glutamine synthetase by adenylylation/deadenylylation.

7.3.4 Biosynthesis of amino acids and its regulation

It is usual to divide the amino acids into biosynthetic families according to their precursors as known from work with *E. coli* and other organisms. Some of these precursors, such as pyruvate, aspartate and glutamate can be formed in phototrophic bacteria by a number of mechanisms, not all of which are found in *E. coli* (see Chapter 6).

As far as the synthesis of the other amino acids from their family precursors is concerned, there are no indications that the biosynthetic routes in phototrophic bacteria are radically different from those in *E. coli*, apart from one known exception, namely the synthesis of isoleucine in *Chlorobium* (Hoare & Gibson, 1964; see Datta, 1978 for review). However, our knowledge is sketchy and new variations will probably be discovered as more pathways and organisms are investigated. Where the phototrophic bacteria do differ from *E. coli* and among themselves, is in the variety of feedback mechanisms involved in amino acid synthesis. In the few purple non-sulphur bacteria that have been examined, several interesting variations of end product inhibition in the aspartate family have been uncovered, mainly by Datta, Gest and their co-workers (Datta & Gest, 1964; Datta *et al.*, 1973). They have shown that aspartate kinase, the first enzyme in the branched pathway responsible for the biosynthesis of lysine, methionine, threonine and isoleucine, is regulated in these bacteria by concerted feedback or compensatory feedback. This is in sharp contrast to the isoenzymic pattern of regulation present in enteric bacteria.

7.4 GENETIC STUDIES OF NITROGEN FIXATION AND AMMONIA ASSIMILATION

Detailed information regarding the genetic regulation of nitrogen fixation requires that the number and arrangement of the genes involved in this process (*nif* genes) be determined. For this purpose many mutants of phototrophic bacteria unable to use molecular nitrogen as sole nitrogen source (Nif^-) have been isolated. Several such mutants of *Rp. capsulata* have been genetically crossed using the 'gene transfer agent' of this bacterium (Wall *et al.*, 1975). This is a phage-like particle which can carry out general transduction in *Rp. capsulata*, and can be used for fine-structure mapping of the genome (Marrs, 1974; see also Chapter 10). Several different Nif^- mutants have been identified on the basis of restoration of Nif^+ character to cells after recombination between different mutant isolates.

The Nif^- mutants tested show no H_2-evolution during phototrophic growth in malate + glutamate medium. This capacity returned to spontaneous revertants and transferants of these mutants. These observations strongly support the suggestion that in phototrophic bacteria, nitrogenase and H_2-evolving hydrogenase activities are catalysed by the same enzyme complex (Wall *et al.*, 1975). The use of H_2 as an electron donor for photoautotrophic growth is mediated by a different hydrogenase system. In accordance with this, the Nif^- mutants grow readily, photosynthetically, with ammonia as the nitrogen source and $H_2 + CO_2$ as the sole source of reducing power and carbon respectively. On the other hand, Nif^- mutations in *Rp. acidophila* affect not only nitrogen fixation but also H_2 oxidation (Siefert & Pfennig, 1978). Hydrogenase (uptake) activities of two mutants were about tenfold less than that of the wild-type grown under similar conditions. Since the genetic defect probably is a single point mutation, it was concluded that a genetic or regulatory linkage exists between nitrogenase and hydrogenase in *Rp. acidophila*. This is in accordance with the suggestion that a physiological role of uptake hydrogenase in N_2 fixing organisms is to recycle H_2 evolved via the nitrogenase.

A mutant of *Rs. rubrum* has been isolated, which shows a nitrogenase-mediated H_2 production about twice that of the wild type strain (Weare, 1978). This mutant exhibits significant derepression of nitrogenase biosynthesis in the presence of ammonia or alanine, and produces only low levels of glutamate synthase. The genetic alteration in glutamate synthase seems to prevent the formation of glutamate in sufficient amounts to provide repressive levels of glutamine via glutamine synthetase.

The glutamate analog, and glutamine synthetase inhibitor L-methionine-DL-sulfphoximine (MSO) (Ronzio & Meister, 1968) causes derepression of nitrogenase biosynthesis in *Rs. rubrum*. In the presence of MSO, this bacterium excretes large amounts of ammonia, fixed from N_2, into the growth-medium (Weare & Shanmugam, 1976). Methionine sulphoximine inhibits glutamine synthetase activity as well as growth of *Rp. capsulata* on either N_2 or ammonia as nitrogen source (Johansson & Gest, 1976). Subsequent addition of glutamine reversed the inhibition of growth. Also in *Rp. capsulata* MSO, and the related compound β-N-oxalyl-L-α,β-diaminopropionic acid, cause derepression of nitrogenase biosynthesis (Meyer & Vignais, 1979). It is probable that products derived from ammonia *via* glutamine synthetase play a role in regulating the biosynthesis of nitrogenase in phototrophic bacteria.

Mutants of *Rs. rubrum* resistant to MSO have been isolated, and these also show nitrogenase-mediated H_2 production in the presence of the inhibitor and high concentrations of ammonia (Johansson, unpublished). Other types of derepressed mutants have been isolated from *Rp. capsulata* (Wall & Gest, 1979). These were isolated as glutamine auxotrophs (Gln$^-$ mutants), in which the phenotype is due to single-site mutations. The different Gln$^-$ mutations were mapped using the 'gene transfer agent', and found to be clustered on the chromosome, possibly representing one or a small number of closely linked genes. One of the mutants had less than 1% of the glutamine synthetase activity of the parental strain, while the level of glutamate synthase was the same in both strains. The results supported the conclusion that *Rp. capsulata* uses the glutamine synthetase/glutamate synthase sequence of reactions as the primary pathway for the assimilation of ammonia, produced from N_2 or added exogenously (Johansson & Gest, 1976). Nitrogen fixation as well as H_2 production of the mutant was fully derepressed in the presence of high concentrations of ammonia. Such genetic disruption of the close coupling between production and biosynthetic utilization of ammonia, may result in continuous conversion of N_2 to ammonia which accumulates in the growth medium. When the *Rp. capsulata* mutant was grown phototrophically on limiting glutamine as nitrogen source and gassed continuously with 5% CO_2 in N_2, the production of extracellular ammonia was exponential during the first 8 hours and continued at a linear rate for 3 days after termination of growth. The maximum rate of ammonia production was 0.6 μmol per h and mg dry weight. Molecular hydrogen could be used as a reductant to some extent, instead of malate, in the conversion of N_2 to ammonia by resting cells of the mutant. The fact that Gln⁻ mutants can be isolated from *Rp. capsulata*, which are derepressed for nitrogenase in the presence of high concentrations of ammonia, indicates that in phototrophic bacteria, as in *Klebsiella pneumoniae* (Streicher et al., 1974), glutamine synthetase and/or other Gln gene products are involved in the regulation of nitrogen fixation.

It can be expected that mutants of the Nif⁻ phenotype will include strains with a variety of genetic defects, for example in structural genes for nitrogenase, electron carriers uniquely associated with N_2 reduction or enzymes of ammonia assimilation. One mutant of *Rp. capsulata* has been isolated, which does not seem to fit into any of these categories. This strain was isolated as a Nif⁻ mutant, but showed a marked pleiotropic defect with respect to nitrogen nutrition (Wall et al., 1977). Thus, in addition to loss of capacity to grow on N_2, it showed little or no ability to use a large number of organic nitrogen sources, that support growth of the wild type parent, in spite of the fact that this was a single site mutation. The mutant, however, grows well with ammonia as nitrogen source when this is supplied at a concentration of 2 mM or more. Functional catabolic enzymes, nitrogenase and ammonia assimilation enzymes were produced by the mutant. It was also found that the K_m (NH_3) of the glutamine synthetase in the mutant did not differ significantly from that of the wild type enzyme. These results, together with the fact that addition of glutamine and/or glutamate does not promote growth of the mutant with utilizable carbon sources such as malate, indicate that there are no apparent defects in the mutant in the biochemical pathway responsible for conversion of inorganic nitrogen to glutamine and glutamate.

One possible explanation for this seemingly anomalous phenotype (Nit⁻) is suggested by considering the pattern of ammonia assimilation in bacteria, characterized by an efficient incorporation of ammonia into glutamine. This is a crucial nitrogen distribution point for production of numerous metabolites through the activity of various specific glutamine amidotransferases (see Fig. 7.2). These enzymes ordinarily use glutamine as the nitrogen donor but in many instances ammonia at relatively high concentrations can substitute for glutamine. Thus, if the glutamine-donor function of such an amidotransferase were specifically inactivated by mutation, the enzyme would perform biosynthetically only when ammonia is provided at appropriately elevated concentrations. Conclusive elucidation of the defect in this mutant, and others of similar phenotype, awaits the appropriate enzymic tests.

As pointed out earlier, studies on the genetic regulation of nitrogenase require that the arrangement of the Nif genes on the chromosome be determined. Such mapping of the 17 Nif genes in *K. pneumoniae* shows them to be clustered in a single large region on the chromosome close to the *his* operon (Mac Neil et al., 1978). By comparison, no data for the Nif genes of phototrophic bacteria are yet available, which would allow a distinction between this arrangement and one with scattered genes. It would be of particular interest to find out whether the Nif genes are clustered in phototrophic bacteria, and if so, whether the gene(s) for the activating enzyme of *Rs. rubrum* nitrogenase is (are) linked to such a cluster.

7.5 CONCLUDING REMARKS

It should be evident from this chapter that understanding of the mechanism and regulation of nitrogen fixation and ammonia assimilation by phototrophic bacteria is now rapidly advancing. Some of the research areas which are being intensely investigated at the present time are: the molecular properties and role of the nitrogenase activating enzyme found in *Rs. rubrum* and the mechanism of the metabolic 'switch-off' effect of nitrogenase; the genetic mapping of the Nif-genes; regulatory properties of glutamine synthetase; and possible involvement of glutamine synthetase in metabolic and genetic control of nitrogenase. A deeper understanding of the intricate molecular details of these processes may be expected to emerge from these and related studies within this decade.

Note added in proof

Evidence for adenylylation of glutamine synthetase in *Chlorobium* has now been obtained (Khanna & Nicholas, 1983).

REFERENCES

ALEF K., ARP D. J. & ZUMFT W. G. (1981a) Nitrogenase switch-off by ammonia in *Rhodopseudomonas palustris*: loss under nitrogen deficiency and independence from the adenylylation state of glutamine synthetase. *Arch. Microbiol.*, **130**, 138–42.

ALEF K., BURKARDT H.-J., HORSTMAN H.-J. & ZUMFT W. G. (1981b) Molecular characterization of glutamine synthetase from the nitrogen-fixing phototrophic bacterium *Rhodopseudomonas palustris*. *Z. Naturforsch.*, **36c**, 246–54.

ALEF K. & ZUMFT W. G. (1981) Regulatory properties of glutamine synthetase from the nitrogen-fixing phototrophic bacterium *Rhodopseudomonas palustris*. *Z. Naturforsch.*, **36c**, 784–9.

ARNON D. I., LOSADA M., NOZAKI M. & TAGAWA K. (1960) Photofixation of nitrogen and photoproduction of hydrogen by thiosulphate during bacterial photosynthesis. *Biochem. J.*, **77**, 23–24P.

BALTSCHEFFSKY H. & NORDLUND S. (1975) Investigations on biological energy conversion. In *Progress Reports 73/74*, p. 283, Swedish Natural Science Research Council, Stockholm.

BIGGINS D. R., KELLY M. & POSTGATE J. R. (1971) Resolution of nitrogenase of *Mycobacterium flavum* 301 into two components and cross reaction with nitrogenase components from other bacteria. *Eur. J. Biochem.*, **20**, 140–3.

Bregoff H. M. & Kamen M. D. (1952) Studies on the metabolism of photosynthetic bacteria XIV. Quantitative relations between malate dissimilation, photoproduction of hydrogen, and nitrogen metabolism in *Rhodospirillum rubrum*. *Arch. Biochem. Biophys.*, **36**, 202–20.

Brown C. M. & Herbert R. A. (1977a) Ammonia assimilation in purple and green sulphur bacteria. *FEMS Microbiol. Lett.*, **1**, 39–42.

Brown C. M. & Herbert R. A. (1977b) Ammonia assimilation in members of the Rhodospirillaceae. *FEMS Microbiol. Lett.*, **1**, 43–46.

Bulen W. A., Le Comte R. C., Burns R. C. & Hinkson J. (1965) Nitrogen fixation studies with aerobic and photosynthetic bacteria. In *Non-Heme Iron Proteins: Role in Energy Conservation*. (Ed. by A. San Pietro), pp. 261–74. The Antioch Press, Yellow Springs, Ohio.

Burns R. C. & Bulen W. A. (1966) A procedure for the preparation of extracts from *Rhodospirillum rubrum* catalyzing N_2 reduction and ATP-dependent H_2 evolution. *Arch. Biochem. Biophys.*, **113**, 461–3.

Carithers R. P., Yoch D. C. & Arnon D. I. (1979) Two forms of nitrogenase from the photosynthetic bacterium *Rhodospirillum rubrum*. *J. Bact.*, **137**, 779–89.

Carnahan J. E., Mortenson L. E., Mower H. F. & Castle J. E. (1960) Nitrogen fixation in cell-free extracts of *Clostridium pasteurianum*. *Biochim. Biophys. Acta*, **44**, 520–35.

Cusanovich M. A. & Edmondson D. E. (1971) The isolation and characterization of *Rhodospirillum rubrum* flavodoxin. *Biochem. Biophys. Res. Commun.*, **45**, 327–36.

Datta P. (1978) Biosynthesis of amino acids. Chapter 41. In *The Photosynthetic Bacteria* (Ed. by R. K. Clayton & W. R. Sistrom) pp. 779–92. Plenum Press, New York and London.

Datta P., Dungan S. M. & Feldberg R. S. (1973) Regulation of amino acid biosynthesis of the aspartate pathway in different micro-organisms. In: *Genetics of Industrial Microorganisms*, Vol. 1 (Ed. by Z. Vanek, Z. Hostalek & J. Cudlin) pp. 177–91. Academia, Prague.

Datta P. & Gest H. (1964) Alternative patterns of end product control in biosynthesis of amino acids of the aspartic family. *Nature Lond.*, **203**, 1259–61.

Davies W. & Ormerod J. G. (1982) Glutamine synthetase in *Chlorobium limicola* and *Rhodospirillum rubrum*. *FEMS Microbiol. Lett.*, **13**, 75–78.

Emerich D. W. & Burris R. H. (1978) Complementary functioning of the component proteins of nitrogenase from several bacteria. *J. Bact.*, **134**, 936–43.

Evans M. C. W., Telfer A. & Smith R. V. (1973) The purification and some properties of the molybdenum-iron protein of *Chromatium* nitrogenase. *Biochim. Biophys. Acta*, **310**, 344–52.

Gest H. (1972) Energy conversion and generation of reducing power in bacterial photosynthesis. *Adv. microb. Physiol.*, **7**, 245–82.

Gest H., Kamen M. D. & Bregoff H. M. (1950) Studies on the metabolism of photosynthetic bacteria. *J. Biol. Chem.*, **182**, 153–70.

Hageman R. V. & Burris R. H. (1980) Electron allocation to alternative substrates of *Azotobacter* nitrogenase is controlled by the electron flux through dinitrogenase. *Biochim. Biophys. Acta*, **591**, 63–75.

Hallenbeck P. C., Meyer C. M. & Vignais P. M. (1982) Nitrogenase from the photosynthetic bacterium *Rhodopseudomonas capsulata*: purification and molecular properties. *J. Bact.*, **149**, 708–17.

Hardy R. W. F. & Burns R. C. (1968) Biological nitrogen fixation. *Ann. Rev. Biochem.*, **37**, 331–58.

Herbert R. A., Siefert E. & Pfennig N. (1978) Nitrogen assimilation in *Rhodopseudomonas acidophila*. *Arch. Microbiol.*, **119**, 1–5.

Hillmer P. & Fahlbusch K. (1979) Evidence for an involvement of glutamine synthetase in regulation of nitrogenase activity in *Rhodopseudomonas capsulata*. *Arch. Microbiol.*, **112**, 213–18.

Hillmer P. & Gest H. (1977) H_2 metabolism in the photosynthetic bacterium *Rhodopseudomonas capsulata*: production and utilization of H_2 by resting cells. *J. Bact.*, **129**, 732–9.

Hoare D. S. & Gibson J. (1964) Photoassimilation of acetate and the biosynthesis of amino acids by *Chlorobium thiosulphatophilum*. *Biochem. J.*, **91**, 546–59.

HONG M. M., SHEN S. C. & BRAUNSTEIN A. E. (1959) Distribution of L-alanine dehydrogenase and L-glutamate dehydrogenase in *Bacilli. Biochim. Biophys. Acta*, **36**, 288–9.

JOHANSSON B. C. & GEST H. (1976) Inorganic nitrogen assimilation by the photosynthetic bacterium *Rhodopseudomonas capsulata. J. Bact.*, **128**, 683–8.

JOHANSSON B. C. & GEST H. (1977) Adenylylation/deadenylylation control of the glutamine synthetase of *Rhodopseudomonas capsulata. Eur. J. Biochem.*, **81**, 365–71.

JONES B. L. & MONTY K. J. (1979) Glutamine as a feedback inhibitor of the *Rhodopseudomonas sphaeroides* nitrogenase system. *J. Bact.*, **139**, 1007–13.

KAMEN M. D. & GEST H. (1949) Evidence for a nitrogenase system in the photosynthetic bacterium *Rhodospirillum rubrum. Science*, **109**, 560.

KIM J. S., ITO K. & TAKAHASHI H. (1980) The relationship between nitrogenase activity and hydrogen evolution in *Rhodopseudomonas palustris. Agric. Biol. Chem.*, **44**, 827–33.

KHANNA S. & NICHOLAS D. J. D. (1983) Some properties of glutamine synthetase and glutamate synthase from *Chlorobium vibrioforme f. thiosulfatophilum. Arch. Microbiol.*, **134**, 98–103.

KLEMME J.-H. (1968) Untersuchungen zur Photoautotrophie mit molekularem Wasserstoff bei neuisolierten schwefelfreien Purpurbakterien. *Arch. Mikrobiol.*, **69**, 29–42.

LEPO J. E., STACEY G., WYSS O. & TABITA F. R. (1979) The purification of glutamine synthetase from *Azotobacter* and other procaryotes by blue sepharose chromatography. *Biochim. Biophys. Acta* **568**, 428–36.

LIND C. J. & WILSON P. W. (1941) Mechanism of biological nitrogen fixation. VIII. Carbon monoxide as an inhibitor for nitrogen fixation by Red Clover. *J. Amer. Chem. Soc.*, **63**, 3511–14.

LINDSTROM E. S., BURRIS R. H. & WILSON, P. W. (1949) Nitrogen fixation by photosynthetic bacteria. *J. Bact.*, **58**, 313–15.

LINDSTROM E. S., LEWIN S. M. & PINSKY M. J. (1951) Nitrogen fixation and hydrogenase in various bacterial species. *J. Bact.*, **61**, 481–7.

LINDSTROM E. S., NEWTON J. W. & WILSON P. W. (1952) The relationship between photosynthesis and nitrogen fixation. *Proc. natn. Acad. Sci. USA*, **38**, 392–6.

LINDSTROM E. S., TOVE S. R. & WILSON P. W. (1950) Nitrogen fixation by the green and purple sulfur bacteria. *Science*, **112**, 197–8.

LUDDEN P. W. & BURRIS R. H. (1976) Activating factor for the iron protein of nitrogenase from *Rhodospirillum rubrum. Science*, **194**, 424–6.

LUDDEN P. W. & BURRIS R. H. (1978) Purification and properties of nitrogenase from *Rhodospirillum rubrum*, and evidence for phosphate, ribose and an adenine-like unit covalently bound to the iron protein. *Biochem. J.*, **175**, 251–9.

LUDDEN P. W. & BURRIS R. H. (1979) Removal of an adenine-like molecule during activation of dinitrogenase reductase from *Rhodospirillum rubrum. Proc. natn. Acad. Sci. USA*, **76**, 6201–5.

LUDDEN P. W. & BURRIS R. H. (1981) *In vivo* and *in vitro* studies on ATP and electron donors to nitrogenase in *Rhodospirillum rubrum. Arch. Microbiol.*, **130**, 155–8.

LUDDEN P. W., OKON Y. & BURRIS R. H. (1978) The nitrogenase system of *Spirillum lipoferum. Biochem. J.*, **173**, 1001–3.

LUDDEN P. W. & PRESTON G. G. (1981) Comparison of active and inactive forms of dinitrogenase reductase from *Rhodospirillum rubrum. Fourth International Symposium on Nitrogen Fixation*, Canberra, Australia, abstract no. 71.

MAC NEIL T., MAC NEIL D., ROBERTS G. P., SUPIANO M. A. & BRILL W. J. (1978) Fine-structure mapping and complementation analysis of *nif* (nitrogen fixation) genes in *Klebsiella pneumoniae. J. Bact.* **136**, 253–66.

MADIGAN M. T., WALL J. D. & GEST H. (1979) Dark anaerobic dinitrogen fixation by a photosynthetic microorganism. *Science*, **204**, 1429–30.

MARRS B. (1974) Genetic recombination in *Rhodopseudomonas capsulata. Proc. natn. Acad. Sci. USA* **71**, 971–3.

MEYER J., KELLY B. C. & VIGNAIS P. M. (1978a) Nitrogen fixation and hydrogen metabolism in photosynthetic bacteria. *Biochimie*, **60**, 245–60.

MEYER J., KELLY B. C. & VIGNAIS P. M. (1978b) Effect of light on nitrogenase function and synthesis in *Rhodopseudomonas capsulata*. *J. Bact.*, **136**, 201–8.
MEYER J. & VIGNAIS P. M. (1979) Effects of L-methionine-DL-sulfoximine and β-N-oxalyl-L-α,β-diaminopropionic acid on nitrogenase biosynthesis and activity in *Rhodopseudomonas capsulata*. *Biochem. Biophys. Res. Commun.*, **89**, 353–9.
MORTENSON L. E. & THORNELEY R. N. F. (1979) Structure and function of nitrogenase. *Ann. Rev. Biochem.*, **48**, 387–418.
MUNSON T. O. & BURRIS R. H. (1969) Nitrogen fixation by *Rhodospirillum rubrum* grown in nitrogen-limited continuous culture. *J. Bact.*, **97**, 1093–8.
MURA U. & STADTMAN E. R. (1981) Glutamine synthetase adenylylation in permeabilized cells of *Escherichia coli*. *J. Biol. Chem.*, **256**, 13014–21.
NEILSON A. H. & DOUDOROFF M. (1973) Ammonia assimilation in blue-green algae. *Arch. Mikrobiol.*, **89**, 15–22.
NEILSON A. H. & NORDLUND S. (1975) Regulation of nitrogenase synthesis in intact cells of *Rhodospirillum rubrum*. Inactivation of nitrogen fixation by ammonia, L-glutamine and L-asparagine. *J. gen. Microbiol.*, **91**, 53–62.
NORDLUND S. & EKLUND R. (1979) Nitrogen fixation in *Rhodospirillum rubrum*—the switch-off effect. Third International Symposium on Photosynthetic Prokaryotes, Oxford, England, abstract no. B 47.
NORDLUND S. & ERIKSSON U. (1979) Nitrogenase from *Rhodospirillum rubrum*: relation between the 'switch-off' effect and the membrane component, hydrogen production and acetylene reduction with different ratios of the nitrogenase components. *Biochim. Biophys. Acta*, **547**, 429–37.
NORDLUND S., ERIKSSON U. & BALTSCHEFFSKY H. (1977) Necessity of a membrane component for nitrogenase activity in *Rhodospirillum rubrum*. *Biochim. Biophys. Acta*, **462**, 187–95.
NORDLUND S., ERIKSSON U. & BALTSCHEFFSKY H. (1978) Properties of the nitrogenase system from a photosynthetic bacterium, *Rhodospirillum rubrum*. *Biochim. Biophys. Acta*, **504**, 248–54.
OLSON J. (1978) Confused story of '*Chloropseudomonas ethylica* 2K'. *Int. J. System. Bacteriol.*, **28**, 128–9.
ORMEROD J. G. & GEST H. (1962) Symposium on metabolism of inorganic compounds IV. Hydrogen photosynthesis and alternative metabolic pathways in photosynthetic bacteria. *Bacteriol. Rev.*, **26**, 51–66.
ORMEROD J. G., ORMEROD K. S. & GEST H. (1961) Light-dependent utilization of organic compounds and photoproduction of molecular hydrogen by photosynthetic bacteria; relationships with nitrogen metabolism. *Arch. Biochem. Biophys.*, **94**, 449–63.
PRATT D. C. & FRENKEL A. W. (1959) Studies on nitrogen fixation and photosynthesis of *Rhodospirillum rubrum*. *Plant Physiol.*, **34**, 333–7.
RONZIO R. A. & MEISTER A. (1968) Phosphorylation of methionine sulfoximine by glutamine synthetase. *Proc. natn. Acad. Sci. USA*, **59**, 164–70.
SAKHNO O. N., IVANOVSKII R. N. & KONDRAT'EVA E. N. (1981) The glutamine synthetase-glutamate synthase system in *Rhodopseudomonas sphaeroides*. *Mikrobiologiya*, **50**, 607–12.
SCHICK H.-J. (1971a) Substrate and light dependent fixation of molecular nitrogen in *Rhodospirillum rubrum*. *Arch. Mikrobiol.*, **75**, 89–101.
SCHICK H.-J. (1971b) Interrelationship of nitrogen fixation, hydrogen evolution and photoreduction in *Rhodospirillum rubrum*. *Arch. Mikrobiol.*, **75**, 102–109.
SCHNEIDER K. C., BRADBEER C., SINGH R. N., WANG L. C., WILSON P. W. & BURRIS R. H. (1960) Nitrogen fixation by cell-free preparations from microorganisms. *Proc. Natn. Acad. Sci. USA*, **46**, 726–33.
SIEFERT E. (1976) Die Fixierung von molekularem Stickstoff bei phototrophen Bacterien am beispiel von *Rhodopseudomonas acidophila*. Ph.D. Thesis, University of Göttingen, West-Germany.

SIEFERT E. & PFENNIG N. (1978) Hydrogen metabolism and nitrogen fixation in wild type and Nif⁻ mutants of *Rhodopseudomonas acidophila*. *Biochimie.*, **60**, 261–5.
SIEFERT E. & PFENNIG N. (1980) Diazotrophic growth of *Rhodopseudomonas acidophila* and *Rhodopseudomonas capsulata* under microaerobic conditions in the dark. *Arch. Microbiol.* **125**, 73–77.
SMITH R. V., TELFER A. & EVANS M. C. W. (1971) Complementary functioning of nitrogenase components from a blue-green alga and a photosynthetic bacterium. *J. Bact.*, **107**, 574–5.
SOLIMAN A., NORDLUND S., JOHANSSON B. C. & BALTSCHEFFSKY, H. (1981) Purification of glutamine synthetase from *Rhodospirillum rubrum* by affinity chromatography. *Acta Chem. Scand.*, **B 35**, 63–64.
STADTMAN E. R. & GINSBURG A. (1974) The glutamine synthetase of *Escherichia coli*: structure and control. In: *The Enzymes*, 3rd Ed. (Ed. by P. Boyer,) p. 755–807. Academic Press, New York.
STADTMAN E. R., GINSBURG A., CIARDI J. E., YEH J., HENNIG S. B. & SHAPIRO B. M. (1970) Multiple molecular forms of glutamine synthetase produced by enzyme catalyzed adenylylation and deadenylylation reactions. *Adv. Enzyme Regul.*, **8**, 99–118.
STREICHER S. L., SHANMUGAM K. T., AUSUBEL F., MORANDI C. & GOLDBERG R. B. (1974) Regulation of nitrogen fixation in *Klebsiella pneumoniae*: evidence for a role of glutamine synthetase as a regulator of nitrogenase synthesis. *J. Bact.*, **120**, 815–21.
SWEET W. J. & BURRIS R. H. (1981) Inhibition of nitrogenase activity by NH_4^+ in *Rhodospirillum rubrum*. *J. Bact.*, **145**, 824–31.
TRONICK S. R., CIARDI J. E. & STADTMAN E. R. (1973) Comparative biochemical and immunological studies of bacterial glutamine synthetases. *J. Bact.*, **115**, 858–68.
TSO M.-Y. (1974) Some properties of the nitrogenase proteins from *Clostridium pasteurianum*. Molecular weight, subunit structure, isoelectric point and EPR spectra. *Arch. Microbiol.*, **99**, 71–80.
TYLER B. (1978) Regulation of the assimilation of nitrogen compounds. *Ann. Rev. Biochem.*, **47**, 1127–62.
VEEGER C., HAAKER H. & LAANE, C. (1981) Energy transduction and nitrogen fixation. In *Current Perspectives in Nitrogen Fixation* (Ed. by A. H. Gibson & W. E. Newton) pp. 101–4, Australian Academy of Science, Canberra.
WALL J. D. & GEST H. (1979) Derepression of nitrogenase activity in glutamine auxotrophs of *Rhodopseudomonas capsulata*. *J. Bact.*, **137**, 1459–63.
WALL J. D., JOHANSSON B. C. & GEST H. (1977) A pleiotropic mutant of *Rhodopseudomonas capsulata* defective in nitrogen metabolism. *Arch. Microbiol.*, **115**, 259–63.
WALL J. D., WEAVER P. F. & GEST H. (1975) Genetic transfer of nitrogenase—hydrogenase activity in *Rhodopseudomonas capsulata*. *Nature Lond.* **258**, 630–1.
WEARE N. M. (1978) The photoproduction of H_2 and NH_4^+ fixed from N_2 by a derepressed mutant of *Rhodospirillum rubrum*. *Biochim. Biophys. Acta*, **502**, 486–94.
WEARE N. M. & SHANMUGAM K. T. (1976) Photoproduction of ammonium ion from N_2 in *Rhodospirillum rubrum*. *Arch. Microbiol.*, **110**, 207–13.
WINTER H. C. & ARNON D. I. (1970) The nitrogen fixation system of photosynthetic bacteria: I. Preparation and properties of a cell-free extract from *Chromatium*. *Biochim. Biophys. Acta*, **197**, 170–9.
WINTER H. C. & OBER J. A. (1973) Isolation of particulate nitrogenase from *Chromatium* strain D. *Plant Cell Physiol.*, **14**, 769–73.
YOCH D. C. (1979) Manganese, an essential trace element for N_2 fixation by *Rhodospirillum rubrum* and *Rhodopseudomonas capsulata*: role in nitrogenase regulation. *J. Bact.*, **140**, 987–95.
YOCH D. C. (1980) Regulation of nitrogenase A and R concentrations in *Rhodopseudomonas capsulata* by glutamine synthetase. *Biochem. J.*, **187**, 273–6.
YOCH D. C. & ARNON D. I. (1970) The nitrogen fixation system of photosynthetic bacteria II.

Chromatium nitrogenase activity linked to photochemically generated assimilatory power. *Biochim. Biophys. Acta* **197**, 180–4.

YOCH D. C. & ARNON D. I. (1975) Comparison of two ferredoxins from *Rhodospirillum rubrum* as electron carriers for the native nitrogenase. *J. Bact.*, **121**, 743–5.

YOCH D. C. & CANTU M. (1980) Changes in the regulatory form of *Rhodospirillum rubrum* nitrogenase as influenced by nutritional and environmental factors. *J. Bact.*, **142**, 899–907.

ZUMFT W. G., ALEF K. & MUMMLER S. (1981) Regulation of nitrogenase activity in *Rhodospirillaceae*. In *Current Perspectives in Nitrogen Fixation* (Ed. by A. M. Gibson & W. E. Newton) pp. 190–3, Australian Academy of Science, Canberra.

ZUMFT W. G. & CASTILLO F. (1978) Regulatory properties of the nitrogenase from *Rhodopseudomonas palustris*. *Arch. Microbiol.*, **117**, 53–60.

ZUMFT W. C. & NORDLUND S. (1981) Stabilization and partial characterization of the activating enzyme for dinitrogenase reductase (Fe-protein) from *Rhodospirillum rubrum*. *FEBS Lett.*, **127**, 79–82.

Chapter 8. Ecology of Phototrophic Bacteria

H. VAN GEMERDEN AND H. H. BEEFTINK

8.1 INTRODUCTION

Phototrophic bacteria often form massive populations in nature. In such blooms, the productivity can be very high. Wetzel (1973) reported the production of 5700 mg C m^{-3} day^{-1} by a *Chromatium* sp. in Smith Hole, a value exceeding by far the algal productivity. In Fayetteville Green Lake, a *Chlorobium* was reported to produce 1600 mg C m^{-3} day^{-1} (Culver & Brunskill, 1969). Usually, however, the production in the bacterial bloom is 100–200 mg C m^{-3} day^{-1}, accounting for about one third of the total photosynthetic production. A compilation of data, collected from various lakes, is given by Biebl & Pfennig (1979) and Wetzel (1975). Blooms strongly facilitate the measurement of the activities of phototrophic organisms in nature.

Furthermore, in anaerobic habitats, phototrophic bacteria play a significant role in the interconnection of the cycles of the two major elements: carbon and sulphur. Fig. 8.1 illustrates the way in which these cycles are related and driven by the input of light energy, captured by the phototrophic partners. In the decay of organic matter, substrates are produced for sulphate-reducing bacteria which in turn provide the anoxygenic phototrophic bacteria with sulphide, used as the electron donor. The formation of the bloom requires an additional input, however, e.g. dead algae or allochtonous sulphide. A detailed description of such an ecosystem is given by Pfennig (1978).

It is difficult to predict the specific occurrence of one or other type of phototrophic bacterium owing to the great metabolic diversity within the four families. The facultatively aerobic purple non-sulphur bacteria, being predominantly photo-organotrophic, may be expected in environments where more or less reduced products accumulate from fermentation processes. The availability of sulphate also creates ideal conditions for sulphate-reducing bacteria, however, and it has been suggested that the development of Rhodospirillaceae is hampered by sulphide, which promotes the growth of Chromatiaceae and Chlorobiaceae (Pfennig, 1967, 1978). Purple non-sulphur bacteria are meta-

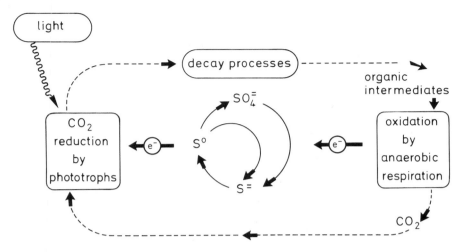

Fig. 8.1 Simplified scheme of the interrelations between the cycles of carbon and sulphur in anaerobic habitats.

bolically versatile and flexible, and thus can be expected and do in fact occur, in a wide variety of natural systems (Kaiser, 1966; Biebl, 1973).

With respect to their capacity to grow either aerobically or photo-organotrophically, the Chloroflexaceae are best characterized as the green counterparts of the Rhodospirillaceae. However, massive developments of Chloroflexaceae have been reported from hot springs. The organisms profit in this habitat from the organic substances provided by cyanobacteria (Pierson & Castenholz, 1974a, b; Pierson, 1973) although they can also grow photo-autotrophically using sulphide as electron donor (Madigan & Brock, 1975). The abundance of these organisms in aquatic systems has been summarized by Padan (1979).

The three main prerequisites for bacterial photosynthesis by Chromatiaceae or Chlorobiaceae are light, absence of oxygen, and the availability of a suitable electron donor, a combination achieved only in aqueous environments such as ponds and lakes. The penetration of light into water, however, results in oxygenic photosynthesis by cyanobacteria and algae. That the bacterial photosynthesis nevertheless occurs can be explained as follows:

1. The primary production by aerobic phototrophs eventually leads (through decay processes) to the availability of organic substances for respiration. At greater depths, and thus lower light intensities, oxygen consumption may exceed oxygen production, eventually resulting in anaerobic conditions.
2. Phototrophic bacteria can photosynthesize at much lower light intensities than green algae (Pfennig & Biebl, 1976; Parsons & Takahashi, 1973).

3. The differences in the near-infrared and visible absorption maxima of eukaryotic algae and anoxygenic phototrophs enable the latter to grow below an algal bloom. However, light at a wavelength over 700 nm is strongly absorbed by water. In deeper waters, therefore, phototrophic bacteria rely on the carotenoid absorption.
4. Organic matter which sediments into the anaerobic parts of the water body, is utilized in fermentation and anaerobic respiration, resulting in sulphide production. In stagnant waters the borderline between aerobic and anaerobic water may be shifted upwards, an effect sometimes enhanced by allochtonous sulphide, e.g. sulphur springs.

The relative importance of these (generally not independent) factors may vary. The simultaneous availability of light, whose intensity decreases with depth, and sulphide, whose concentration decreases towards the surface, may create growth conditions for Chromatiaceae and Chlorobiaceae somewhere along these gradients. This could explain the mass developments at specific depths, probably enhanced by the simultaneous availability of simple organic compounds. In addition, the oxygen profile is important. The environment is thus adequately described as a multi-gradient habitat, emphasizing that the conditions vary strongly with depth. This existence 'on the slopes', into which phototrophic sulphur bacteria are forced is further complicated by diurnal fluctuations caused by the fact that some gradient-determining factors are light-dependent, whereas others are not, e.g. oxygen production and oxygen consumption, respectively. Needless to say, such fluctuating gradients have important implications for the organisms involved. In the remainder of this chapter it becomes evident that there are considerable gaps in our knowledge of the abilities of phototrophic bacteria to respond to such environmental conditions.

Questions on the quantitative role of organisms in nature, and the factors that determine their activity require a bilateral approach. Field studies on phototrophic bacteria are discussed in Section 8.2; Section 8.3 deals with experiments under controlled laboratory conditions.

8.2 FIELD STUDIES

In the following, attention is focussed on stratified lakes, which are the most extensively studied habitats of phototrophic bacteria. The circumstances in other habitats, to be mentioned briefly in Section 8.2.2, are comparable to some extent to the bloom-forming conditions encountered in stratified lakes.

Often light, sulphide, and oxygen, and sometimes predation are mentioned as the decisive factors for the development of blooms. However, with few exceptions these statements are based on observations made in enrichment or pure cultures. Very few investigations have actually been performed on blooms, and some of these are discussed below.

8.2.1 Stratified lakes

Physical and chemical characterization

Many lakes, in particular deep ones, show temperature stratification: a warmer upper layer floats upon a colder bottom layer. The development of stratification is visualized in Fig. 8.2. At the end of the cold season, the water is uniform with respect to temperature and oxygen concentration (Fig. 8.2a). Spring illumination causes the water temperature to rise. As a result of circulation, caused by wind, the upper layer remains homogeneous with respect to density, temperature, and oxygen concentration. This well-mixed, illuminated, warmer upper layer is called the epilimnion, and because of the density difference it does not easily mix with the colder and darker hypolimnion. The metalimnion is the interface between these two layers, usually showing a very steep temperature gradient.

In the hypolimnion of oligotrophic lakes, i.e. with low nutrient concentrations, the low respiration rate results in a so-called orthograde oxygen profile (Fig. 8.2b). In eutrophic, i.e. nutrient-rich lakes, however, the higher respiration causes oxygen depletion of the hypolimnion, resulting in a clinograde oxygen profile (Fig. 8.2c). The latter situation is not restricted to inland lakes, but is also observed in anoxic fjords.

Density stratification, as described above, is often not stable throughout the year. The start of the cold season results in the cooling of the surface water, thus reducing the density differences. Eventually, winds mix the epilimnion with the hypolimnion. If this turn-over involves the whole body of water, the lake is called holomictic. In meromictic lakes, a deep water layer called the

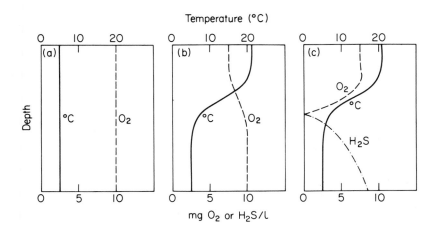

Fig. 8.2 Development of stratification in oligotrophic and eutrophic lakes. a: situation at the end of the cold season; b: summer stratification in an oligotrophic lake; c: summer stratification in an eutrophic lake. See text for further explanation.

monimolimnion, is prevented from mixing owing to high concentrations of solutes which result in high specific densities. Such conditions are often found in lakes with a bottom layer of relict sea water or in lakes which are connected to the sea by a small inlet (Goehle & Storr, 1978; Indrebø et al., 1979a, b).

For further details of terms and definitions, the reader is referred to handbooks on limnology, e.g. Wetzel, 1975; Hutchinson, 1957; Ruttner, 1962.

Blooms of phototrophic bacteria

In the anaerobic parts of stratified eutrophic lakes conditions are favourable for sulphate-reducing bacteria, and considerable sulphide accumulation may occur (up to 20 mmol/l, Chebotarev et al., 1974). The concentration of sulphide decreases towards the surface and often the upper limit coincides with the lower limit of oxygen.

Phototrophic bacteria bloom most profusely at or just below the aerobic/anaerobic transition zone, where the light intensities are usually low. Generally Chromatiaceae or Chlorobiaceae are the organisms involved; Rhodospirillaceae have been reported only as negligible sub-populations (Takahashi & Ichimura, 1968; Biebl, 1973; Gorlenko, 1978; Kuznetzov & Gorlenko, 1973) and the same applies to Chloroflexaceae in lakes (Dubinina & Kuznetzov, 1976).

Frequently, mixed populations of Chromatiaceae and Chlorobiaceae are observed. In a homogeneous habitat, this may point to the coexistence of these bacteria. Although mechanisms are known that would enable such a coexistence (Fredrickson, 1977; see also Section 8.3), it should once more be stressed that the blooms are formed along a complicated environmental gradient, and that conditions may change for each centimetre of depth (see Jørgensen et al., 1979).

It is of interest that Chromatiaceae usually proliferate somewhat higher up than Chlorobiaceae (Anagnostidis & Overbeck, 1966; Caldwell, 1977; Caldwell & Tiedje, 1975; Gorlenko & Kuznetzov, 1972; Kuznetzov & Gorlenko, 1973). Such distribution patterns suggest the existence of different niches along the gradients of sulphide, light, oxygen, and temperature.

The influence of light intensity

It has often been argued that the reduced light intensities in stratified lakes limit the growth rate of phototrophic bacteria. Studies with this in mind have been elegantly performed by Sorokin (1970). ^{14}C-carbonate was added to samples of the bloom of *Chromatium* in Lake Belovod, followed by incubation in transparent vials at various depths. It was found that ^{14}C-incorporation was strongly dependent on depth, and thus on light intensity. The *Chromatium* population appeared to have a 25-fold over-capacity for photosynthesis. This value is an overestimate, since samples incubated closer to the surface are subject to increased temperature as well as increased light intensity. However, correction of the original data, arbitrarily assuming a Q_{10}-value of 2, does not affect the original conclusion that the organisms in the bloom were light limited.

Using the same method, light limitation was demonstrated for a bloom of *Chlorobium phaeovibrioides* in Lake Repnoe (Gorlenko *et al.*, 1973) and of *Chlorobium phaeobacteroides* in Lake Kinneret (Bergstein *et al.*, 1979). Although in the latter case, samples from the bloom showed only slightly increased rates of CO_2 fixation when incubated at depths corresponding to a 5-fold higher light intensity as compared to that in the bloom, the authors pointed out that a daily wind creates an internal seiche in the lake with an amplitude of 8 m. As a consequence, the organisms are periodically irradiated with an intensity at which increased rates of CO_2 fixation could indeed be observed.

Vertical migration of phototrophic bacteria

Since gradients in stratified lakes may show diurnal fluctuations, it would be a competitive advantage for organisms to maintain their relative position in the shifting gradients by some sort of motility.

Chlorobiaceae are non-motile but presumably the gas vacuoles, observed in several species, have a buoyancy-regulating function, similar to that of the cyanobacteria (Clark & Walsby, 1978a, b).

Many of the Chromatiaceae possess flagella and are highly motile. Sorokin

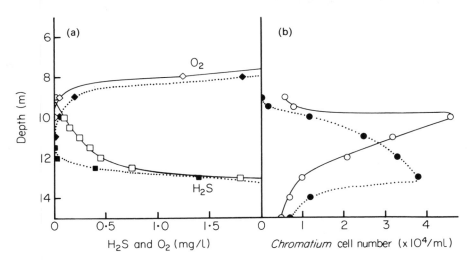

Fig. 8.3 Stratification in Lake Belovod. a: daily fluctuations of the oxygen and sulphide profiles; b: vertical migration of the *Chromatium* bloom. Open symbols: 6 a.m., closed symbols: 12 a.m. Redrawn after Sorokin (1970).

(1970) reported diurnal fluctuations of a bloom of *Chromatium* in Lake Belovod. The vertical migration, in which the organisms seemed to move at a speed of $100\,\mu m\,s^{-1}$, coincided with the fluctuations of the sulphide and oxygen gradients (Fig. 8.3). In contrast, Caldwell & Tiedje (1975) were unable to demonstrate vertical migration in Wintergreen Lake and expressed doubt as to whether phototrophic bacteria are able to develop enough speed, referring to experiments of Vaituzis & Doetsch (1969), who estimated the speed of purple sulphur bacteria to be $10\text{--}30\ \mu m\,s^{-1}$.

Predation

Planktonic predators are often found just above blooms of phototrophic bacteria and many authors have reported on the predation of phototrophic bacteria by zooplanktonic species (Culver & Brunskill, 1969; Northcote & Halsey, 1969; Goulder, 1971, 1972, 1974; Jørgensen & Fenchel, 1974; Sorokin & Donato, 1975; Hayden, 1972a, b; Takahashi & Ichimura, 1968). Usually, the predators are various kinds of Copepoda and Cladocera and their intestines are coloured as a result of the ingestion of the phototrophic bacteria thus aiding detection of the phenomenon. Sometimes, bacterial pigments are even observable in a secondary predator (Goehle & Storr, 1978).

Generally, zooplanktonic predators are obligate aerobes and it is usually supposed that they dive periodically into the anaerobic layer inhabited by their prey organisms (Caldwell, 1977).

To our knowledge, the only quantitative field study on the effect of grazing on a bloom of phototrophic bacteria has been performed by Sorokin (1970). In Lake Belovod, a heavy bacterial bloom is observed at the aerobic/anaerobic interface (Fig. 8.4a). Numerically, the phototrophic bacteria are of minor importance but on a biomass basis, the large purple sulphur bacterium (*Chromatium okenii*, Pfennig, 1978) contributes substantially (Fig. 8.4b). In the aerobic part of the lake, just above the bacterial bloom, large numbers of red-coloured Cladocera were found (Sorokin, 1970). Samples from different depths were incubated either untreated or after filtration to remove predators, in transparent vials at the same depths. Total bacterial counts were made at both the start and end of the incubation period. From the published data for filtered samples, the average specific growth rate of the bacteria can be calculated, assuming negligible substrate depletion, lysis, parasitism, and sedimentation during the incubation period. The data from unfiltered samples allow the estimation of the effect of predation, expressed as a negative specific growth rate of the bacteria (Fig. 8.4c).

The production of cell material is a function of specific growth rate (Fig. 8.4c) and bacterial biomass (Fig. 8.4b); the same holds in a negative way for predation. The calculated average production and predation are compared in Fig. 8.4d. It is obvious that in the Cladocera bloom there is a net production deficit, whereas the anaerobic bacterial bloom shows a net production surplus.

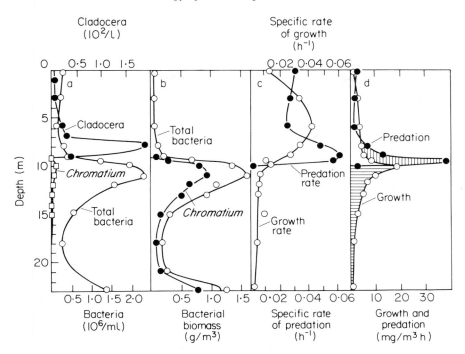

Fig. 8.4 Predation in Lake Belovod. a: vertical distribution of predators and prey organisms; b: total bacterial wet weight and calculated biomass of the purple sulphur bacteria; c: specific growth rate at various depths and specific rate of predation expressed as the negative growth rate of the prey organisms; d: gross bacterial production and predation: horizontal hatching indicates a production surplus, vertical hatching a production deficit. Cell numbers and total biomass data from Sorokin (1970).

If these data reflected the actual situation, the bacterial population at 9 m would decrease rapidly, whereas that at 11 m would increase. The actual, stable situation is explained by the fact that predation at 9 m is balanced by production at 11 m. This is due to the predator diving although turbulence may also have contributed to it (Sorokin, 1970). Sorokin's data do not enable us to discriminate between predation on *Chromatium* and other, smaller, bacteria. Assuming that the predators have no preference, it can be deduced that the phototrophic bacterium is growing at an average specific growth rate of approximately $0.012\ h^{-1}$. Any preference of the predators for the phototrophs would require faster growth of *Chromatium* to explain the observations.

These data show unequivocally that purple sulphur bacteria can contribute significantly to secondary production processes in stratified lakes. It is likely that the growth rates of phototrophic organisms are determined by these phenomena, in addition to such parameters as substrate concentration and light intensity.

8.2.2 Other habitats

In addition to their occurrence in stratified lakes, phototrophic bacteria have been observed in high numbers in other habitats with light, sulphide, oxygen, and redox gradients, e.g. shallow forest ponds sheltered from wind by trees. Heavy clouds of swarming cells of large type *Chromatium* have been reported from such habitats (Schlegel & Pfennig, 1961). The organisms were observed to rise from the mud in the early morning, and disappear again later in the day, conceivably hiding from the increased oxygen concentration, a result of algal photosynthesis.

In other shallow bodies of water, which are almost entirely aerobic, phototrophic bacteria are forced to reside permanently in the uppermost layers of the sediment. There, conditions are still anaerobic and some light penetrates. This situation is frequent in marine sediments (Fenchel, 1969; Jørgensen & Fenchel, 1974; Jørgensen, 1977; Potts & Whitton, 1979). A very pronounced example of such a sediment ecosystem is the so-called 'Farbstreifensandwatt' (a coloured laminated ecosystem). Here, on a millimetre scale, dense populations of different phototrophic bacteria overly each other, with layers of cyanobacteria and diatoms on top (Schulz, 1936; Schulz & Meyer, 1939; Hoffmann, 1942; Hanselmann, 1980; Rongen & Stal, personal communication).

Phototrophic bacteria can also occur in high numbers in the sediments of rice fields, where they contribute significantly to nitrogen fixation (Kobayashi & Hague, 1971; Watanabe & Furusaka, 1980). The predominant occurrence of Rhodospirillaceae in such sediments is of particular interest, and has been explained on the basis of the profuse availability of organic compounds (Okuda *et al.*, 1957). The phototrophs cannot usually degrade macromolecules themselves, and depend on the hydrolysing activities of other heterotrophic organisms (Pfennig, 1978). Rhodospirillaceae are also frequently encountered in sewage lagoons (Cooper, 1962; Siefert *et al.*, 1978).

Environments that are extreme with respect to salt content and temperature (hot sulphur springs, desert lakes) can also be inhabited by phototrophic bacteria (Pierson, 1973; Pierson & Castenholz, 1974a, b; Jannasch, 1957; Imhoff *et al.*, 1978, 1979; Butlin & Postgate, 1954; Cohen *et al.*, 1977). Often, the dominating forms are gliding green bacteria (Chloroflexaceae) and halophilic purple sulphur bacteria (*Ectothiorhodospira halophila*).

Finally, phototrophic bacteria have been found associated with eukaryotes such as marine sponges (Imhoff & Trüper, 1976).

8.3 EXPERIMENTAL ECOLOGY OF PHOTOTROPHIC BACTERIA

The somewhat confusing term 'experimental ecology' refers to laboratory experiments which may contribute to the understanding of natural ecosystems.

As a rule, organisms are isolated from nature and studied under laboratory conditions in batch and/or continuous culture with the aim of extrapolating the results to the organism's habitat. Such an approach is chosen because natural ecosystems are usually far too complicated to enable the elucidation of the actual events and inter-relationships by direct field measurements.

However, reservations should be exercised in the extrapolation of laboratory data to nature. For instance, population interactions may easily be under-estimated. Also, steady-state continuous culture data must be interpreted carefully. Characteristically, continuous cultures are homogeneous, and once they reach steady state are (theoretically) stable. Nature, on the other hand, is continually shifting from one transient state to another, providing fluctuating growth conditions all the time, and being in addition spatially heterogeneous. However, continuous cultures are elegant and indispensable devices in the study of growth at the low nutrient concentrations commonly found in nature.

8.3.1 Population interactions

In natural environments populations rarely subsist independently. As a rule, the component populations of an ecosystem are significantly influenced by each other's activities.

It is common practice to name the type of interaction, but this does not mean that rigid distinctions exist between the different types of interaction. General schemes have been proposed for binary population interactions. Unfortunately, none is generally accepted. Therefore, the examples of binary-population interactions discussed in this section are preceded by a short description. In nature, interactions are not restricted to just two populations and furthermore they usually change with time.

Mutualism

An association of two different organisms having an obligate reliance on each other for life in the ecosystem. The association is beneficial to both. Often referred to as symbiosis (Alexander, 1971; Meers, 1973). When the two species have complementary metabolic patterns, also referred to as syntrophy (Pfennig, 1978).

The extremely low total inorganic sulphur content observed particularly in freshwater habitats towards the end of the summer (Stuiver, 1967; van Gemerden, 1967), may cause mutualism between sulphate-reducing and sulphide-oxidizing bacteria.

In the laboratory, it has been demonstrated with mixed populations of *Desulfovibrio desulfuricans* and *Chromatium vinosum*, that growth of both can occur with no other sulphur compounds than those present in the trace element solution (van Gemerden, 1967). The medium was otherwise complete and contained formate as electron donor for *Desulfovibrio*.

Had *Chlorobium* been the phototrophic partner instead of *Chromatium*, the oxidation of sulphide could have resulted in the formation of extracellular sulphur as an intermediate. If the sulphide-producing bacterium were able to utilize sulphur as electron acceptor in addition to sulphate, a competition for sulphur between the two organisms would take place. However, both sulphur and sulphate can be recycled in this system. For the growth of either organism it is actually irrelevant which sulphur compound is utilized by *Desulfovibrio*.

A more intriguing case of mutualism plus competition occurs when the heterotrophic organism is unable to reduce sulphate, and thus completely depends for its growth on the production of sulphur by the phototrophic bacterium. Such an organism is *Desulfuromonas acetoxidans*, which was grown in mixed culture with *Chlorobium limicola* by Biebl & Pfennig (1978). For good development of the mixed cultures only catalytic amounts of sulphur were required. Therefore, it was deduced that *Desulfuromonas* successfully competed with *Chlorobium* for extracellular sulphur. This is beneficial to both organisms, since sulphate cannot be recycled in this system. In these experiments, acetate served as the electron donor for the heterotrophic organism. Provided that sulphide is present, acetate can also be assimilated by *Chlorobium* (Sadler &

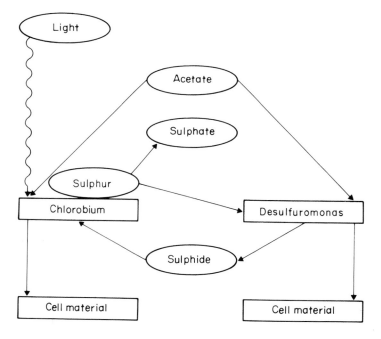

Fig. 8.5 Schematic representation of interactions between *Chlorobium* and *Desulfuromonas* with acetate as the organic substrate.

Stanier, 1960). Therefore, it is unknown to what extent acetate is utilized by either organism (Fig. 8.5).

The relations become even more complicated when mixed cultures of *Chlorobium* and *Desulfovibrio* are grown on ethanol (Biebl & Pfennig, 1978). *Chlorobium* is unable to assimilate ethanol. Therefore, in such a system *Chlorobium* is growing on sulphide and acetate, both produced by *Desulfovibrio* growing on ethanol, provided that sulphate is regenerated through sulphide oxidation by *Chlorobium* (Fig. 8.6).

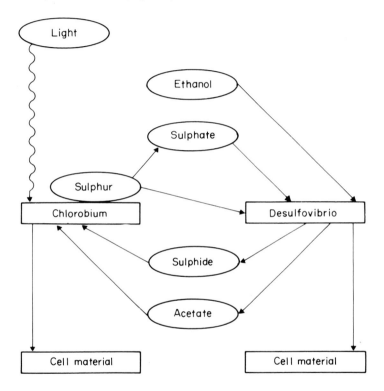

Fig. 8.6 Schematic representation of the interactions between *Chlorobium* and *Desulfovibrio* with ethanol as the organic substrate.

In a three-membered system, studied by Matheron & Baulaigue (1976), *Chlorobium* grew on sulphide and acetate, both produced by *Desulfovibrio*. The latter organism used as substrates the organic compounds produced by *Escherichia coli* in the fermentation of glucose: the ultimate substrate.

Competition

An interaction between two populations consuming the same nutrient, whose concentration has a rate-limiting effect on either one (weak competition) or both populations (strong competition). Also referred to as resource-type competition (Fredrickson, 1977). The limiting factor can also be space, light, etc.

In batch cultures, rate-limiting nutrient concentrations occur only towards the end of the logarithmic growth phase. The time span of such low nutrient concentrations can in theory be extended indefinitely by applying continuous culture techniques (see Herbert et al., 1956; Tempest, 1970; Veldkamp, 1976). In such systems, steady states develop after a certain period of time. The specific growth rate μ then equals the dilution rate D, and the substrate concentration s equals the value predicted by the Monod equation

$$\mu = \mu_{max} \, s/(K_s + s) \qquad \text{Eqn. 8.1}$$

In this equation, K_s is the saturation constant, numerically equal to the substrate concentration at which $\mu = \frac{1}{2}\mu_{max}$.

In continuous culture of a mixed population, the organism requiring the lowest concentration of the limiting substrate to grow at a specific rate equal to the dilution rate, outcompetes all other organisms*. The other organisms still grow, but at a lower specific rate than the dilution rate, and hence they are eventually washed out.

According to Eqn. 8.1, organisms show a saturation-type of growth-rate response to increasing substrate concentrations. In the case of crossing μ/s curves, the outcome of the competition depends on the substrate concentration, and thus on the dilution rate.

One example of such a resource-type competition among phototrophic bacteria has been worked out theoretically by Veldkamp & Jannasch (1972) on the basis of an unintentional selection experiment reported by van Gemerden and Jannasch (1971). Upon inoculation of a continuous culture (D = 0.04 h^{-1}) with a supposedly pure culture of a marine *Thiocystis violacea*, the colour changed from purple to brown; this was found to be due to the growth of a *Chromatium vinosum*. The *Thiocystis* had been maintained in batch culture by repeated feeding with sulphide. The *Chromatium* was studied in detail and found to be very sensitive to raised sulphide concentrations (van Gemerden & Jannasch, 1971).

Veldkamp and Jannasch (1972) offered as the most likely explanation for the phenomena observed, that the two organisms had crossing μ/sulphide curves.

---

* Usually, it is said that the winning organism has a better affinity for the substrate, and often this is judged from the magnitude of the K_s values. However, the specific growth rate at the prevailing substrate concentration is the determinative factor. A mere comparison of K_s values makes sense only for organisms with similar μ_{max} values (Zevenboom, 1980; see also Healey, 1980; Brown & Molot, 1980). When substrate concentrations are low compared with the K_s values, the ratio μ_{max}/K_s, being the initial slope of the Monod curve, can conveniently be used to compare the affinities of different organisms for a common substrate.

Later work (van Gemerden, unpublished) demonstrated this idea to be correct. However, the fact that in the batch cultures sulphide was added intermittently may have contributed to the sustained presence of the *Chromatium* in the *Thiocystis* culture (see Section 8.3.2.).

Predation

An interaction which is beneficial to one organism (the predator), whereas the other (the prey) is harmed, and possibly eliminated. The predator is usually large compared to the prey, and may literally swallow the latter.

Laboratory studies in which phototrophic bacteria are used as prey organisms, are not often performed, despite the conceivably significant contribution of these bacteria to the food of predators (see also Section 8.2.1).

Gophen *et al.*, (1974) measured the uptake of bacteria and algae by *Ceriodaphnia reticulata*, the most common herbivorous zooplankton species in Lake Kinneret. In this lake, a bloom of the brown *Chlorobium phaeobacteroides* develops in the summer, and large numbers of Cladocera and Copepoda were observed just above the layer of bacteria. The estimation of a possible food preference of *Ceriodaphnia* was studied by offering the predator ^{3}H-labelled *Chlorobium* and/or ^{14}C-labelled algae. The presence of algae did not greatly affect the uptake of bacteria. However, the uptake of algae was reduced drastically in the presence of *Chlorobium*. These data indicate a preference of *Ceriodaphnia* for bacteria over algae, and point to the importance of the brown sulphur bacterium as a food source for secondary producers in Lake Kinneret.

8.3.2 Coexistence of competing species

Some interactions, if exerted for a sufficiently long time in homogeneous continuous flow systems, may result in the complete exclusion of one of the organisms. In nature this probably never happens because environmental conditions are always changing to become more favourable for other species. Moreover, wash-out does not occur in natural systems but to some extent it simulates predation. In addition, nature, by virtue of its heterogeneity, provides 'hiding places'. In this respect, a continuous culture with wall growth (a well-known spurious phenomenon) is to be regarded a better imitation of nature than one without.

However, coexistence is not necessarily based upon nature's spatial heterogeneity, as is illustrated below.

Dual substrate limitation

Stable coexistence of two (or more) competing species is possible under nutrient-limited growth conditions provided two (or more) substrates are available simultaneously. Coexistence under these conditions has not only been predicted on theoretical grounds (Yoon *et al.*, 1977; Phillips, 1973; Taylor & LeB. Williams, 1975), but also observed experimentally (Laanbroek *et al.*, 1979; Gottschal *et al.*, 1979). Coexistence is conceivable when substrate 1 is exclusively utilized by organism A, compensating in this way for its lower growth-rate response on substrate 2 as compared to organism B. The utilization of both substrates by both organisms may also result in coexistence provided the organisms have complementary growth-rate responses on the substrates. Such interrelations are also found among the phototrophic bacteria. Two examples of special interest are discussed below.

Rhodopseudomonas capsulata/Chromatium vinosum Surprisingly at first sight, the purple non-sulphur bacterium *Rp. capsulata* not only grows fast in sulphide/CO_2 media, but also has a very high affinity for sulphide (Hansen, 1974; Dijkstra *et al.*, 1983). If the kinetic parameters for growth of this organism on sulphide are compared to those of *Chr. vinosum* the latter organism would be expected to be outcompeted at all non-inhibitory concentrations of sulphide (Fig. 8.7).

However, *Rp. capsulata* is unable to oxidize sulphide any further than elemental sulphur which is deposited outside the cell (Hansen, 1974).

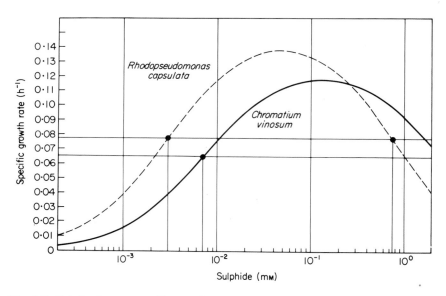

Fig. 8.7 Relation between specific growth rate μ and sulphide concentration for *Rhodopseudomonas capsulata* and *Chromatium vinosum*. Sulphide is toxic. Therefore, unlike the saturation-type of response described by the Monod equation, the growth rate decreases at high sulfide concentrations. The inhibition constant K_i can be compared with the saturation constant K_s.

Chromatium, on the other hand, is able to utilize extracellular sulphur as electron donor. Thus, theoretically, coexistence is to be expected in sulphide-limited chemostats. In this case the second substrate (extracellular sulphur) is produced from the first (sulphide) by one member of the community (*Rp. capsulata*) and exclusively utilized by the other (*Chr. vinosum*), whereas the best affinity for the first substrate is found in the organism which is unable to oxidize the second substrate. And indeed, balanced coexistence was observed in such systems (Wijbenga, unpublished).

In the oxidation of sulphide to sulphur and sulphur to sulphate, two and six electrons are released, respectively. The yield per unit of reducing power was reported to be almost identical for the two organisms (van Gemerden & Beeftink, 1978; Dijkstra *et al*., 1983), and thus the biomass ratio can be used directly to calculate the amount of substrate utilized by each of the competing populations.

During the stable coexistence on sulphide, virtually all the sulphide was oxidized to sulphate. *Rp. capsulata* contributed only about 5% to the total biomass indicating that *Chr. vinosum* must have oxidized most of the sulphide and all of the sulphur (both intracellular and extracellular) to sulphate.

Other Rhodospirillaceae, (e.g. *Rp. palustris*) can oxidize sulphide to sulphate, without the formation of detectable amounts of sulphur (Hansen, 1974). It is not known at present whether such organisms can successfully compete with *Chromatium* species on the basis of their sulphide affinity.

In acetate-limited continuous cultures, inoculated with both *Rp. capsulata* and *Chr. vinosum*, the latter was pushed back, slowly, but definitely, which points to a better affinity of *Rp. capsulata* for acetate. Mixotrophic conditions (sulphide/CO_2 + acetate) again resulted in coexistence. In this case *Rp. capsulata* contributed approximately 15% to the total biomass (Wijbenga, unpublished) showing that the simultaneous presence of acetate improves the competitive position of *Rp. capsulata*. However, the fact that *Rp. capsulata* does not dominate in spite of superior affinities for both sulphide and acetate, emphasizes the severe drawback of not being able to oxidize elemental sulphur. The poor ability to handle elevated sulphide concentrations (see *Fluctuating Substrate Concentrations*, p. 162) is another trait that may explain why these organisms never bloom in nature.

Chromatium vinosum/Chlorobium limicola In growing the purple sulphur bacterium *Chr. vinosum* and the green sulphur bacterium *Chl. limicola* in mixed continuous cultures with sulphide as the sole electron donor in the reservoir solution, stable coexistence was observed over a wide range of dilution rates (Table 8.1).

Again, the second substrate (extracellular sulphur) is produced from the first (sulphide) by one member of the community (*Chlorobium*). In this case, however, both organisms are able to utilize both substrates, and therefore the fate of sulphide and extracellular sulphur cannot as easily be traced back as in

Table 8.1 *Chromatium vinosum* and *Chlorobium limicola* in balanced coexistence in sulphide-limited continuous cultures at various dilution rates. Data from van Gemerden & Beeftink (1981).

Dilution rate (h^{-1})	*Chromatium*		*Chlorobium*	
	protein (%)	biovolume (%)	protein (%)	biovolume (%)
0.02	11	10	89	90
0.05	9	13	91	87
0.08	51	50	49	50
0.10	98	95	2	5

Protein data were calculated on the basis of the bacteriochlorophyll *a* and *c* content. Biovolumes were estimated with electronic sizing equipment (Coulter Counter).

the preceding example. At the lower dilution rates, *Chlorobium* was found to dominate by 90%. In view of the fact that *Chl. limicola* has a far better affinity for sulphide than *Chr. vinosum*, it was concluded that *Chlorobium* must have oxidized a substantial amount of the extracellular sulphur in addition to most of the sulphide (van Gemerden & Beeftink, 1981).

Fluctuating substrate concentrations

Rhodopseudomonas capsulata/Chromatium vinosum Studies with *Rp. capsulata* revealed that in continuous cultures sulphide and acetate are assimilated simultaneously (Dijkstra *et al.*, 1983). However, acetate-grown cells were found to have a low potential for sulphide oxidation. Sulphide had to be introduced into heterotrophically growing cultures with great care; otherwise it accumulated and wash-out followed (Wijbenga & van Gemerden, 1981). It was suggested that reduced levels of Calvin-cycle enzymes exhibited during heterotrophic growth (see Chapter 6) were responsible for the low potential rates of sulphide oxidation.

In mixed cultures of *Rp. capsulata* and *Chr. vinosum*, exposed to similar fluctuations, the *Chromatium* effectively prevented the accumulation of sulphide, and thus allowed the sustained presence of its competitor. However, the result was an even stronger dominance of *Chromatium* than described under 'Dual substrate limitation'.

Chromatium vinosum/Chromatium weissei In general, the Monod equation adequately describes the relation between concentration of limiting nutrient and specific growth rate. It must be stressed, however, that this refers to balanced growth only. Instantaneously increasing growth rates in response to increased substrate concentrations have been reported for *E. coli* (Mateles *et al.*, 1965), but were not observed in *Chr. vinosum* (Beeftink & van

Gemerden, 1979). However, *Chromatium* does start to assimilate substrates immediately after being exposed to light (Trüper & Schlegel, 1964; van Gemerden, 1968a).

The importance of such phenomena in the competition between phototrophic bacteria was illustrated by laboratory experiments performed with 'large' and 'small' representatives of the Chromatiaceae (van Gemerden, 1974).

The large *Chromatium* species are intriguing creatures, which bloom frequently. Yet, samples from such blooms, taken to the laboratory, and incubated with sulphide in the light, are rapidly overgrown by smaller species. This led to the speculation that the two forms might have crossing μ/sulphide curves. However, for *Chr. vinosum* and *Chr. weissei*, representing the small and the large forms respectively, this was found not to be the case: the small species grew faster at all sulphide concentrations.

Subsequently, experiments were performed to estimate the maximal rates of substrate utilization under conditions of unbalanced growth. Pure cultures

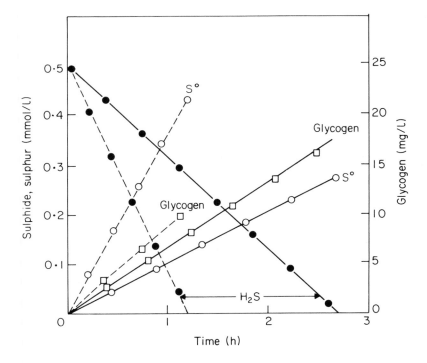

Fig. 8.8 Sulphide oxidation and product formation in non-growing cultures of *Chromatium vinosum* (full lines) and *Chromatium weissei* (broken lines). Data recalculated on a standard cell density of 10 mg cell nitrogen/l (van Gemerden, 1974; supplemented with unpublished glycogen data).

were fed with sulphide in the dark. Upon re-illumination, no growth was observed for some hours, but sulphide was oxidized and concomitantly, glycogen was deposited inside the cells. The data (Fig. 8.8) show that the specific rate of sulphide oxidation is much higher in the *Chr. weissei* culture.

In natural ecosystems, the concentration of sulphide shows diurnal fluctuations. This was simulated in the laboratory by exposing a sulphide-limited continuous culture inoculated with both *Chr. vinosum* and *Chr. weissei*, alternately to light and dark periods. Under these conditions, the two organisms were found to coexist. The explanation offered was that most of the sulphide that accumulated in the dark periods was rapidly oxidized by *Chr. weissei* (compare Fig. 8.8), whereas the extremely low sulphide concentrations during the remainder of the light periods were advantageous for *Chr. vinosum*. Longer dark periods (prolonged sulphide accumulation) indeed resulted in increased *Chr. weissei* populations.

Note that the specific rate of CO_2 fixation exhibited by *Chr. weissei* ($0.055\ h^{-1}$) is higher than its maximum specific growth rate ($0.040\ h^{-1}$). This implies that even if sulphide were present at elevated concentrations for a prolonged period of time, this would only be advantageous for a limited period. Thus, a high uptake rate is of particular importance if the concentration of substrates fluctuates in the organism's habitat.

Chromatium/Chlorobium With respect to the competition between purple and green sulphur bacteria, the temporary absence of sulphide is of particular relevance. *Chlorobium* species lack assimilatory sulphate reduction, grow slowly with sulphur as electron donor, and require sulphide for the incorporation of acetate (Sadler & Stanier, 1960). In addition to all this, Chlorobiaceae are obligate anaerobes, in contrast to Chromatiaceae (Kämpf & Pfennig, 1980). In natural ecosystems, absence of sulphide often coincides with the presence of oxygen. Fluctuations in the concentration of sulphide and in particular the absence of sulphide, can thus be expected to be very disadvantageous for *Chlorobium*.

In gradient ecosystems, green sulphur bacteria often develop below the purple sulphur bacteria. It could very well be that the vertical distribution of green and purple sulphur bacteria is governed by the presence of oxygen, rather than by the availability of sulphide. Conceivably, the purple sulphur bacteria prevent the penetration of oxygen to deeper layers, in this way enabling their potential competitors to develop below them. In this respect, the lower light requirements of the green sulphur bacteria are very relevant (see also 'Dual Wave length utilization' and Section 8.3.3).

Dual wavelength utilization

Blooms of phototrophic bacteria are frequently encountered at depths of 10–15 m. Although light requirements of phototrophic bacteria are in general much lower than those of, e.g. algae (Parsons & Takahashi, 1973), organisms flourishing at such depths are often light-limited (see also Section 8.2.1).

In lake Kinneret, a small population of *Thiocapsa roseopersicina* was observed to coexist with the dominating *Chlorobium phaeobacteroides* (Bergstein *et al.*, 1979). Exploratory experiments with the two Kinneret isolates (Veldhuis, unpublished), in which the full incandescent spectrum was applied, showed the *Thiocapsa* to have higher light requirements than the *Chlorobium*, as has been reported for other purple and green sulphur bacteria (Biebl & Pfennig, 1978). However, not only the quantity of light, but also its quality changes with depth (see Pfennig, 1967). In general, only light between 400 and 600 nm reaches the blooms. If absorption by the organisms differs in this spectral region, coexistence becomes feasible in much the same way as with dual substrate limitation. This has been worked out theoretically for two algae by Stephanopoulos and Fredrickson (1979). Accordingly, the ratio between the two phototrophic bacteria in Lake Kinneret can be expected to depend on the spectral composition of the light, and thus on the depth. The *in vivo* spectra of the Kinneret isolates did indeed show differences in the 400– 600 nm region, as has also been reported for the organisms blooming in Lake Faro (Trüper & Genovese, 1968). Actual data on the vertical distribution of the *Chlorobium* and the *Thiocapsa* are lacking, however.

In this respect, the experiments of Matheron and Baulaigue (1977) are of interest. They incubated mixed pure cultures of *Thiocystis violacea* and *Chlorobium phaeovibrioides* at various depths. The purple sulphur bacterium developed well only in bottles incubated near the surface; in all other bottles the brown *Chlorobium* was enriched.

Phototrophic organisms generally increase their content of light-harvesting moieties in response to sub-saturating light intensities. Increased bacteriochlorophyll contents have often been reported (see Chapter 2). Both natural blooms and the isolated dominant species have been reported to show increased specific contents of both bacteriochlorophyll and carotenoids under light-limiting conditions (Abellà *et al.*, 1980; Guerrero *et al.*, 1980). Göbel (1978) observed an increase in the carotenoid content of *Rp. capsulata* growing in reduced monochromatic light (522 nm) as compared to growth under light saturation.

8.3.3 Maintenance and survival

Even when not growing, organisms require energy, primarily for the maintenance of a proton motive force (pmf) over the membrane (Konings & Veldkamp, 1980). This pmf is needed for osmotic regulation, control of internal pH, synthesis of ATP, and turnover of essential macromolecules (Pirt, 1975; Tempest, 1970; Dawes, 1976). The energy required for all processes not directly related to growth, including that for motility, is collectively called the energy of maintenance. As a consequence, curves describing the relation between specific growth rate and the concentration of the limiting energy source do not go through the origin and the ordinate intercept of the extrapolated curve has been described as the specific maintenance rate μ_e (Powell, 1967). The same principle applies to the growth-rate dependency of phototrophic bacteria on the intensity of light.

Various phototrophic bacteria are able to perform fermentation, aerobic and/or anaerobic respiration in addition to photosynthesis. Usually, the prime concern has been the ability of the organisms to grow under such conditions. Although growth indicates that the population contained viable cells, absence of growth certainly does not mean that all the cells are dead. Therefore, growth is not the proper parameter to judge the importance of a process for maintenance.

Fermentation

Fermentation of, e.g. pyruvate or fructose has been reported for several species of the Rhodospirillaceae (see Gürgün *et al.*, 1976; see Chapter 5). As a rule, growth is slow under these conditions: doubling times exceeding 100 hours ($\mu < 0.007\,h^{-1}$) are no exception.

Dark fermentation of exogenous substrates not resulting in growth has also been observed in *Chr. vinosum* (Hendley, 1955; Gürgün *et al.*, 1976). Growth has been reported to occur under these conditions in other purple sulphur bacteria (Krasil'nikova *et al.*, 1976; 1977).

In *Chlorobium*, endogenous reserve materials are broken down in the dark to organic acids. Presumably, this process provides the organisms with energy for maintenance (Sirevåg, 1975; Sirevåg & Ormerod, 1977; see also Section 6.4).

Sulphur respiration

In *Chr. vinosum*, glycogen synthesized in the light in sulphide/CO_2 media is not fermented, but converted to poly-β-hydroxybutyrate and CO_2 concomitantly with the reduction of previously deposited sulphur to sulphide, a process likely to yield some ATP (van Gemerden, 1968a; b). Again, no growth was detected. This process has been described as an endogenous anaerobic respiration (Stanier *et al.*, 1976). The production of sulphide in the dark under similar conditions has been reported for a variety of phototrophic bacteria (Roelofsen, 1935; Larsen, 1953; van Niel, 1936; Hendley, 1955; Trüper & Schlegel, 1964; Trüper & Pfennig, 1966; Oren & Shilo, 1979).

Aerobic respiration

Most Rhodospirillaceae can grow chemo-organotrophically, either under microaerophilic conditions or fully aerobically. The *Chromatiaceae* have long been considered obligate phototrophs. However, dark aerobic growth was reported for *Thiocapsa roseopersicina* Bogorov (1974) and *Amoebobacter roseus* Gorlenko (1974). Recent data conclusively demonstrate that many purple sulphur bacteria can grow chemotrophically at low oxygen tension (Kämpf & Pfennig, 1980). Again, the rate of growth was low compared to that under anaerobic conditions in the light (see Chapter 5). Kämpf & Pfennig (1980) also examined various *Chlorobiaceae* for chemotrophic growth: all were found to be obligate phototrophs.

Low irradiance

Viability studies at low light intensities have been performed with *Chr. vinosum*. The relation between specific growth rate and light intensity was estimated, and, by extrapolation to zero growth rate, the light intensity required for maintenance purposes was deduced. Indeed, at this light intensity *Chromatium* maintained 100% viability for over a week without detectable growth. In the corresponding dark cultures, which were also in nutrient medium, viability decreased rapidly (van Gemerden, 1980).

In washed suspensions of phototrophically grown *Rhodospirillum rubrum* illumination had a marked sparing effect on the degradation of cellular carbohydrates and prolonged the viability (Breznak *et al.*, 1978).

Present data indicate that green sulphur bacteria require less light for maintenance than purple sulphur bacteria (Biebl & Pfennig, 1978; van Gemerden, 1980; Raven & Beardall, 1981). The authors last mentioned have postulated that the low maintenance requirement of the Chlorobiaceae is related not only to the quality and quantity of their photopigments, but also to the spatial distribution of these within the cell.

Summarizing this section, it may be concluded that in phototrophic bacteria fermentation, anaerobic respiration, and low irradiance function primarily, to provide the organisms with energy for maintenance. Aerobic respiration is far more effective, but under conditions where sulphide is respired phototrophic bacteria live in competition with chemolithotropic organisms such as *Thiobacillus*. Although these usually grow much faster than phototrophic organisms under these conditions, this should not prematurely be taken to be the decisive factor for the outcome of this competition. The affinities of the phototrophic bacteria for sulphide under microaerophilic conditions remain to be established and compared to those of the *Thiobacillus* species before predictions in this respect can be made. Nevertheless, in artificial sulphureta, Fenchel (1969) observed the replacement of *Thiobacillus*-like organisms by *Chromatium*-like bacteria, and Jørgensen and Fenchel (1974) reported the appearance of *Chromatium* to coincide with the disappearance of *Beggiatoa*. However, it is not known by which type of metabolism the phototrophic bacteria were growing.

8.3.4 Enrichment cultures

In addition to answering the question of how to obtain a large population of a desired micro-organism, enrichment cultures provide useful information on the relevant parameters that permit an organism to bloom in nature.

The most common procedure is to choose conditions that are selective, so that the desired organism grows faster than the other organisms present. This is the crucial feature of enrichments, and the conditions are not necessarily optimal for growth of the desired organism.

Enrichment techniques can be divided into those based on spatial heterogeneity, and those offering homogeneous growth conditions. Examples of heterogeneous systems are percolating soil columns, Winogradsky columns, and gel-stabilized gradient systems. The wide variety of conditions they offer enable growth of organisms whose nutritional requirements are not (exactly) known. Homogeneous systems, on the other hand, are easier to analyse. Both types have been used for the enrichment of phototrophic bacteria.

Mud columns

The application of mud columns for the enrichment of phototrophic bacteria was originally described by Winogradsky (1888); later modifications were reported by Schrammeck (1934) and Pfennig (Pfennig & Schlegel, 1960; Schlegel & Pfennig, 1961). The system mimics natural habitats. Sulphide produced by sulphate-reducing bacteria—at the expense of, e.g. lower fatty acids which in turn are produced from added cellulose etc., by other heterotrophic bacteria—slowly diffuses into the water overlying a mud layer in which organic matter and sulphate are included. When the column is incubated in daylight, coloured spots become visible after some time, followed by the development of phototrophic bacteria (green and purple sulphur bacteria) in the water column. Without further precautions, fast growing green sulphur bacteria almost invariably outcompete the purple sulphur bacteria. The latter are enriched selectively when red light (>800 nm) is applied. Under such conditions high light intensity and temperature favour the small type of Chromatiaceae, lower light and temperature favour the large forms (Pfennig, 1965).

Diffusion gradients

Parameters such as light intensity, redox potential, sulphide concentration etc., are often mentioned to explain the distribution of various types of phototrophic bacteria in nature. The recognition that these parameters all vary as a function of depth has led Caldwell (Caldwell *et al.*, 1973; Caldwell & Hirsch, 1973) to develop a two-dimensional diffusion gradient which has been used successfully both in the field and in the laboratory. Using gradients of methylamine and sulphide, Caldwell and Hirsch (1973) showed that specific requirements resulted in the separation of *Hyphomicrobium, Rhodomicrobium* and *Thiopedia*. Using gradients of acetate and growth factors occurring in mud, a large variety of microbial colonies developed.

Batch cultures

Organisms growing in batch culture, are not facing a shortage of nutrients and therefore grow at the maximal rate attainable under such conditions. One thus selects for organisms with a high μ_{max}, which might have been numerically in the minority at the start.

Studies of different types of purple and green sulphur bacteria, led van Niel (1931) to formulate conditions for the specific enrichment. By varying the initial pH and sulphide concentration he was able to distinguish between three groups:

1. the small *Chromatium* type;
2. the *Thiocystis* type; and
3. the *Chlorobium* type.

Van Niel used highly selective media in which the smaller forms invariably overgrew the large forms. Careful observation and more precise knowledge of the organisms' nutritional requirements (Pfennig & Schlegel, 1960; Schlegel & Pfennig, 1961), led to a medium for cultivation of the large forms as well (Pfennig, 1961; 1962; 1965). Among the selective parameters were the presence of vitamin B_{12}, the intensity and the regimen of light, and the temperature, in addition to the concentration of sulphide and the pH. Conditions under which the various species may be enriched are listed in Table 8.2. Species with gas vacuoles float to the surface, and thus are easily separated from the others, especially at lower temperatures (Pfennig, 1967).

In general, for the enrichment of the purple and green sulphur bacteria advantage is taken of their ability to grow photolithoautotrophically. Although some of the purple non-sulphur bacteria can oxidize sulphide (Hansen, 1974; Hansen & van Gemerden, 1972), this group is predominantly photo-organotrophic and sulphide usually prevents their growth. The Rhodospirillaceae assimilate a wide variety of organic compounds, although some substrate specificity can be recognized. Most have vitamin requirements. Enrichment media for Rhodospirillaceae thus fulfil the nutritional requirements of sulphate-reducing bacteria as well. Therefore, care has to be taken to eliminate sulphate, which otherwise, through reduction to sulphide, prevents the growth of the purple non-sulphur bacteria.

Continuous cultures

In contrast to batch-culture enrichments, continuous flow systems inoculated with mixed populations favour the growth of those organisms with a strong

Table 8.2 Enrichment conditions for purple sulphur bacteria and green sulphur bacteria (Compiled from Pfennig 1965 = 1, 1967 = 2, 1978 = 3).

Light intensity		Light quality during initial transfers		Light regimen L–D		Initial sulphide concentration		pH		Temp.		Genera, species	
Lux	Ref.	nm	Ref.	h–h	Ref.	mM	Ref.		Ref.	°C	Ref.		Ref.
1000–2000	3	>800	1	24–0	1	2–8	3	6.6–7.2	1	25–30	1	*Chromatium* (small)	1
700–2000	1		2			3	1					*Thiocystis*	3
												Thiocapsa	
												Thiosarcina	
												Ectothiorhodospira	
700–2000	1	day	1	day	1	4–8	3	6.6–7.0	1	25–30	1	*Chlorobium limicola*	3
700–1500	3	incand.	1	24–0	1	3	1					*Prosthecochloris*	3
												Chlorobium phaeobacteroides	
												Chlorobium phaeovibrioides	
300–700	1	>800	1	16–8	1	3	1	6.8–7.2	1	20–25	1	*Chromatium warmingii*	1
			2	day	1								
100–300	1	>800	1	16–8	1	0.4–1	3	6.5–6.8	1	15–20	1	*Chr. okenii, Chr. weissei,*	1
												Thiodictyon	
			2	day	1					15–20	1	*Thiospirillum jenense*	1
100–300	1	>800	1,2	16–8/day	1	<3		6.5–6.8	1	4–10	2	*Amoebobacter, Lamprocystis*	1
200–500	1	day	1	day	1			6.6–7.0	1	16–22	1	*Pelodictyon*	1
		incand.	1	16–8	1								
50–100	3					0.4–2	3			10–20	3	*Pelodictyon*	3
												Ancalochloris	3

affinity for the limiting substrate, rather than those with a high μ_{max}. From an ecological point of view, enrichment in such open systems is more attractive than in closed (batch) systems since the former simulate the low substrate concentrations usually encountered in nature without producing the thin cultures that would result from low substrate concentrations in batch cultures (Veldkamp & Jannasch, 1972).

Usually, one single nutrient dispersed homogeneously determines the outcome of the enrichment. Dual-substrate limitation may result in coexisting populations (see Section 8.3.2). Linked continuous cultures with an inlet of substrate 1 at one end of the series and of substrate 2 at the other have been designed by Cooper and Copeland (1973) and Lovitt and Wimpenny (1979; 1981). In such systems, different ratios of substrates 1 and 2 are obtained in the different cultures. These designs can be used not only to study pure cultures in opposing gradients, but also to enrich for organisms whose quantitative nutritional requirements are not known.

8.4 CONCLUDING REMARKS

> 'Ecological studies aim at supplying the data on which to base an interpretation of the manner in which environmental factors determine the outcome of the struggle for existence' (van Niel, 1955).

Both field studies and laboratory experiments contribute to the understanding of the inter-relationships between organisms and their habitat. Field studies have the advantage that the actual situation is directly involved. However, interpretation is often hampered by the complexity of the ecosystem. Also, environmental factors are not easily controlled by the investigator. In very steep gradients even the precise collection of samples may be difficult. In standard-type laboratory experiments, on the other hand, sampling and the control of the various factors are easy. However, without exception, laboratory studies are oversimplified compared to natural habitats and, although the outcome may correlate well with observations in natural environments, either ignorance or courage is required to extrapolate the data (without reluctance) to the situation in the field. Nevertheless, both field and laboratory observations point to the relevance of light, oxygen, nutrient concentration, microbial interaction and survival characteristics during starvation.

Evidently, our knowledge is far from complete. Both laboratory experiments and field observations can be criticized on a number of points. The fact that some of these criticisms are mentioned below is intended to stimulate future research and certainly not to minimize the importance of work done already.

With respect to the laboratory experiments there are objections related to the test organisms on the one hand and to growth conditions on the other. For example, it is often too easily assumed that the test organism chosen is a fair representative of the group considered to be important, and not a metabolic exception of some sort. Sometimes an organism is even selected for research because it lacks a peculiarity considered to be nasty, e.g. a tendency to flocculate, or to adhere to glass. Frequently the test organism was isolated long ago but is tacitly assumed not to have changed its properties over the years. Also, the number of test organisms grown in mixed culture is often only two, sometimes three, and their behaviour is compared with that of pure cultures. The fact is often neglected—as in our own work—that the growth-rate response of a bacterium to substrate concentration or light intensity may change as the result of the presence of another factor. For example, the μ/light curve of a phototrophic bacterium is not necessarily the same in media containing acetate or acetate plus sulphide. Finally, conditions should be chosen such that an organism can express its full metabolic potential; in this respect the chemotrophic growth of Chromatiaceae must be included in competition studies.

With respect to growth conditions, the habitat concept generally adopted is far too simple to allow extrapolation of the data to natural environments. First, and possibly most important, the organisms are usually cultured in well-stirred homogeneous media because it is not often recognized that ecosystems are spatially structured. More attention should be paid to gradients of sulphide, light, oxygen, redox potential, etc., (Padan, 1979; Wimpenny, 1981). The number of substrates generally involved is small (often only one) and in many cases only the effect of a continuous supply of nutrients is studied, thus favouring organisms with effective metabolic control. For example, the more 'effective' metabolic regulation of *Rp. capsulata* compared to *Chr. vinosum* is a drawback rather than an advantage under fluctuating conditions (see Section 8.3.2).

In growing phototrophic bacteria for competition studies one generally employs the optimal conditions of pH, temperature, etc., rather than simulating natural conditions. This applies also to light quality: usually the full incandescent spectrum is used, although we know that the organisms in deep lakes rely on the light absorbed by their carotenoids.

With respect to field measurements, van Niel's (1955) statement is still valid:

> 'Unfortunately, the relationships between the characteristics of an environment and the flora and fauna found therein must often be deduced from observations made at the time when the organisms are already present in large numbers. This is not always a satisfactory guide to an interpretation of ecological factors, because at such a time the environment may have been considerably modified by the activities of the organisms themselves'.

In addition, data collected while a bloom is present are often limited in number. They may well confirm that there is a bloom, and allow speculation on the significance of certain environmental factors, but usually the data do little to help elucidate the kinetics of the ecosystem. The data collected by Sorokin (1970) on 29th July 1964 demonstrate how much can be learned from an intensive sampling programme.

Actually, 'improvements' can be made rather easily without losing the advantages of a certain approach. In the laboratory, more attention should be paid to the behaviour of mixed cultures; the seven-membered microbial community described by Senior *et al.*, (1976) may serve as an example. In addition, fluctuating concentrations and opposing gradients of mixed substrates (Wimpenny, 1981; Caldwell *et al.*, 1973) deserve our full attention. The stepwise gradients created in the 'gradostat' (Lovitt & Wimpenny 1979, 1981) are to be preferred because they facilitate accurate sampling by normal laboratory routines.

With respect to field work, the estimation of more parameters in samples taken far more frequently may be expected to reveal the significance of (other) factors that can subsequently be studied in detail in the laboratory under controlled conditions. The study performed by Jørgensen *et al.*, (1979) in the Solar Lake may serve as a shining example.

Finally, model systems with various degrees of complexity (Suckow & Schwartz, 1963; Jørgensen & Fenchel, 1974) should be undertaken more often to fill the gap between laboratory experiments and natural ecosystems.

REFERENCES

ABELLÀ C., MONTESINOS E. & GUERRERO R. (1980) Field studies on the competition between purple and green sulfur bacteria for available light (Lake Siso, Spain). *Dev. Hydrobiol.*, **3**, 173–81.

ALEXANDER M. (1971) *Microbial Ecology*, pp. 511. John Wiley & Sons, New York.

ANAGNOSTIDIS K. & OVERBECK J. (1966) Methanoxydierer und hypolimnische Schwefelbakterien. Studien zur ökologischen Biocönotik der Gewassermikroorganismen. *Ber. Dtsch. Bot. Ges.*, **79**, 163–74.

BEEFTINK H. H. & VAN GEMERDEN H. (1979) Actual and potential rates of substrate oxidation and product formation in continuous cultures of *Chromatium vinosum*. *Arch. Microbiol.*, **121**, 161–7.

BERGSTEIN T., HENIS Y. & CAVARI B. Z. (1979) Investigations on the photosynthetic sulfur bacterium *Chlorobium phaeobacteroides* causing seasonal blooms in Lake Kinneret. *Can. J. Microbiol.*, **25**, 999–1007.

BIEBL H. (1973) Die Verbreitung der schwefelfreien Purpurbakterien im Plussee und anderen Seen Ostholsteins. Ph.D. Thesis, University of Freiburg (F.R.G.).

BIEBL H. & PFENNIG N. (1978) Growth yields of green sulfur bacteria in mixed culture with sulfur and sulfate reducing bacteria. *Arch. Microbiol.*, **117**, 9–16.

BIEBL H. & PFENNIG N. (1979) Anaerobic CO_2 uptake by phototrophic bacteria. A Review. *Arch. Hydrobiol. Beih.*, (Ergebn. Limnol.) **12**, 48–58.

BOGOROV L. V. (1974) About the properties of *Thiocapsa roseopersicina* strain BBS, isolated from the estuary of the White Sea. *Microbiology*, **43**, 275–80.

BREZNAK J. A., POTRIKUS C. J., PFENNIG N. & ENSIGN J. C. (1978) Viability and endogenous substrates used during starvation survival of *Rhodospirillum rubrum*. *J. Bacteriol.*, **134**, 381–8.

BROWN E. J. & MOLOT L. A. (1980) Competition for phosphorus among microplankton. *Abstr. Int. Symp. Micr. Ecol.*, **2**, 163.

BUTLIN K. R. & POSTGATE J. R. (1954) The microbiological formation of sulphur in Cyrenaican lakes. In *Biology of Deserts* (Ed. by J. L. Cloudsley-Thompson), pp. 112–22. Institute of Biology, London.

CALDWELL D. E. (1977) The planktonic microflora of lakes. *CRC Crit. Revs. Microbiol.*, **5**, 305–70.

CALDWELL D. E. & HIRSCH P. (1973) Growth of microorganisms in two-dimensional steady-state diffusion gradients. *Can. J. Microbiol.* **19**, 53–8.

CALDWELL D. E., LAI S. H. & TIEDJE J. M. (1973) A two-dimensional steady-state diffusion gradient for ecological studies. *Bull. Ecol. Res. Comm.*, **17**, 151–8.

CALDWELL D. E. & TIEDJE J. M. (1975) The structure of anaerobic bacterial communities in the hypolimnia of several Michigan lakes. *Can. J. Microbiol.*, **21**, 377–85.

CHEBOTAREV E. N., GORLENKO V. M. & KACHALKIN V. I. (1974) Microbial production of hydrogen sulfide in Lake Veisovoe (Slavyansk Lakes). *Microbiology*, **43**, 271–4.

CLARK A. E. & WALSBY A. E. (1978a) The occurrence of gas-vacuolate bacteria in lakes. *Arch. Microbiol.*, **118**, 223–8.

CLARK A. E. & WALSBY A. E. (1978b) The development and vertical distribution of populations of gas-vacuolate bacteria in a eutrophic, monomictic lake. *Arch. Microbiol.*, **118**, 229–33.

COHEN Y., KRUMBEIN W. E. & SHILO M. (1977) Solar lake (Sinai). 2. Distribution of photosynthetic microorganisms and primary production. *Limnol. Oceanogr.*, **22**, 609–20.

COOPER D. C. & COPELAND B. J. (1973) Responses of continuous-series estuarine microecosystems to point-source input variations. *Ecol. Monogr.*, **43**, 213–36.

COOPER R. C. (1962) Photosynthetic bacteria in waste treatment. *Dev. Ind. Microbiol.*, **4**, 95–103.

CULVER D. A. & BRUNSKILL G. J. (1969) Fayetteville Green Lake, New York. V. Studies of primary production and zooplankton in a meromictic lake. *Limnol. Oceanogr.*, **14**, 862–73.

DAWES E. A. (1976) Endogenous metabolism and the survival of starved prokaryotes. *Symp. Soc. Gen. Microbiol.*, **26**, 19–53.

DIJKSTRA A., WIJBENGA D.-J. & VAN GEMERDEN H. (1983) Growth yields of *Rhodopseudomonas capsulata* under autotrophic, heterotrophic and mixotrophic conditions: Anaerobic yields and affinities for sulfide and acetate. *Arch. Microbiol.*, (in press).

DUBININA G. A. & KUZNETZOV S. I. (1976) The ecological and morphological characteristics of microorganisms in Lesnaya Lamba (Karelia). *Int. Rev. Ges. Hydrobiol.*, **61**, 1–19.

FENCHEL T. (1969) The ecology of marine microbenthos. IV. Structure and function of the benthic ecosystem. *Ophelia*, **6**, 1–182.

FREDRICKSON A. G. (1977) Behaviour of mixed cultures of microorganisms. *Ann. Rev. Microbiol.*, **31**, 63–87.

GEMERDEN H. VAN (1967) On the bacterial sulfur cycle of inland waters. Ph.D. Thesis, University of Leiden, The Netherlands.

GEMERDEN H. VAN (1968a) Utilization of reducing power in growing cultures of *Chromatium*. *Arch. Mikrobiol.*, **64**, 111–17.

GEMERDEN H. VAN (1968b) On the ATP generation by *Chromatium* in darkness. *Arch. Mikrobiol.*, **64**, 118–24.

GEMERDEN H. VAN (1974) Coexistence of organisms competing for the same substrate: an example among the purple sulfur bacteria. *Micr. Ecol.*, **1**, 104–19.

GEMERDEN H. VAN (1980) Survival of *Chromatium vinosum* at low light intentsities. *Arch. Microbiol.*, **125**, 115–21.

GEMERDEN H. VAN & BEEFTINK H. H. (1978) Specific rates of substrate oxidation and product formation in autotrophically growing *Chromatium vinosum* cultures. *Arch. Microbiol.*, **119**, 135–43.

GEMERDEN H. VAN & BEEFTINK H. H. (1981) Coexistence of *Chlorobium* and *Chromatium* in a sulfide-limited chemostat. *Arch. Microbiol.*, **129**, 32–4.

GEMERDEN H. VAN & JANNASCH H. W. (1971) Continuous culture of Thiorhodaceae. Sulfide and sulfur limited growth of *Chromatium vinosum*. *Arch. Mikrobiol.*, **79**, 345–53.
GÖBEL F. (1978) Quantum efficiencies of growth. In *The Photosynthetic Bacteria* (Ed. by R. K. Clayton & W. R. Sistrom), pp. 907–25. Plenum Press, New York.
GOEHLE K. H. & STORR J. F. (1978) Biological layering resulting from extreme meromictic stability, Devil's Hole, Abaco Island, Bahamas. *Verh. Internat. Verein. Limnol.*, **20**, 550–5.
GOPHEN M., CAVARI B. Z. & BERMAN T. (1974) Zooplankton feeding on differentially labelled algae and bacteria. *Nature, Lond.*, **247**, 393–4.
GORLENKO V. M. (1974) The oxidation of thiosulfate by *Amoebobacter roseus* in the dark under microaerophilic conditions. *Microbiology*, **43**, 624–5.
GORLENKO V. M. (1978) Phototrophic sulfur bacteria of salt meromictic lakes and their role in the sulfur cycle. In *Environmental Biochemistry and Geomicrobiology*. Vol. I. The aquatic environment (Ed. by W. E. Krumbein), pp. 109–19. Ann Arbor Science, Ann Arbor.
GORLENKO V. M., CHEBOTAREV, E. N. & KACHALKIN, V. I. (1973) Microbiological processes of oxidation of hydrogen sulfide in the Repnoe Lake (Slavonic lakes). *Microbiology*, **42**, 643–6.
GORLENKO V. M. & KUZNETZOW S. I. (1972) Über die photosynthetisierender Bakterien des Kononjer-Sees. *Arch. Hydrobiol.*, **70**, 1–13.
GOTTSCHAL J. C., DE VRIES S. & KUENEN J. G. (1979) Competition between a facultatively chemolithotrophic *Thiobacillus* A2, an obligately chemolithotrophic *Thiobacillus*, and a heterotrophic spirillum for inorganic and organic substrates. *Arch. Microbiol.*, **121**, 241–9.
GOULDER R. (1971) The effects of saprobic conditions on some ciliated Protozoa in the benthos and hypolimnion of a eutrophic pond. *Freshwater Biol.*, **1**, 307–18.
GOULDER R. (1972) Grazing by the ciliated Protozoan *Loxodus magnus* on the alga *Scenedesmus* in a eutrophic pond. *Oikos*, **23**, 109–15.
GOULDER R. (1974) The seasonal and spatial distribution of some benthic ciliated Protozoa in Esthwaite Water. *Freshwater Biol.*, **4**, 127–47.
GUERRERO R., MONTESINOS E., ESTEV I. & ABELLÀ C. (1980) Physiological adaptation and growth of purple and green sulfur bacteria in a meromictic lake (Vila) as compared to a holomictic lake (Siso). *Dev. Hydrobiol.*, **3**, 161–71.
GÜRGÜN V., KIRCHNER G. & PFENNIG N. (1976) Vergärung von Pyruvat durch sieben Arten phototropher Purpurbakterien. *Z. Allg. Mikrobiol.*, **16**, 573–86.
HANSELMANN K. W. (1980) Phototrophic bacteria in stratified sand cores from salt marsh sediments. *Abstr. Int. Symp. Micr. Ecol.*, **2**, 201.
HANSEN T. A. (1974) Sulfide als electronendonor voor Rhodospirillaceae. Ph.D. Thesis, University of Groningen, The Netherlands.
HANSEN T. A. & VAN GEMERDEN H. (1972) Sulfide utilization by purple non-sulfur bacteria. *Arch. Mikrobiol.*, **86**, 49–56.
HAYDEN J. F. (1972a) A limnological investigation of a meromictic lake (Medicine Lake, South Dakota). Masters Thesis, University of South Dakota, USA.
HAYDEN J. F. (1972b) The relationship between zooplankton distribution and physicochemical characteristics in Medicine Lake, South Dakota. *Proc. S. Dak. Acad. Sci.*, **51**, 269.
HEALEY F. P. (1980) Slope of the Monod equation as an indicator of advantage in nutrient competition. *Micr. Ecol.*, **5**, 281–6.
HENDLEY D. D. (1955) Endogenous fermentation in Thiorhodaceae. *J. Bact.*, **70**, 625–34.
HERBERT D., ELSWORTH R. & TELLING R. C. (1956) The continuous culture of bacteria; a theoretical and experimental study. *J. gen. Microbiol.*, **14**, 601–22.
HOFFMANN C. (1942) Beiträge zur Vegetation des Farbstreifensandwattes. *Kieler Meeresforsch.*, **4**, 85–108.
HUTCHINSON G. E. (1957) *A Treatise on Limnology*. Vol. I. pp. 1015. John Wiley & Sons, New York.
IMHOFF J. F., HASHWA F. & TRÜPER H. G. (1978) Isolation of extremely halophilic phototrophic bacteria from the alkaline Wadi Natrun, Egypt. *Arch. Hydrobiol.*, **84**, 381–8.

IMHOFF J. F., SAHL H. G., SOLIMAN G. S. H. & TRÜPER H. G. (1979) The Wadi Natrun: chemical composition and microbial mass developments in alkaline brines of eutrophic desert lakes. *Geomicrobiol. J.*, **1**, 219–34.

IMHOFF J. F. & TRÜPER H. G. (1976) Marine sponges as habitats of anaerobic phototrophic bacteria. *Micr. Ecol.*, **3**, 1–9.

INDREBØ G., PENGERUD B. & DUNDAS I. (1979a) Microbial activities in a permanently stratified estuary. I. Primary production and sulfate reduction. *Mar. Biol.*, **51**, 295–304.

INDREBØ G., PENGERUD B. & DUNDAS I. (1979b) Microbial activities in a permanently stratified estuary. II. Microbial activities in the oxic-anoxic interface. *Mar. Biol.*, **51**, 305–9.

JANNASCH H. W. (1957) Die bakterielle Rotfärbung der Salzseen des Wadi Natrun (Aegypten). *Arch. Hydrobiol.*, **53**, 425–33.

JØRGENSEN B. B. (1977) The sulfur cycle of a coastal marine sediment. *Limnol. Oceanogr.*, **22**, 814–32.

JØRGENSEN B. B. & FENCHEL T. (1974) The sulfur cycle of a marine sediment model system. *Mar. Biol.*, **24**, 189–201.

JØRGENSEN B. B., KUENEN, J. G. & COHEN, Y. (1979) Microbial transformations of sulfur compounds in a stratified lake (Solar Lake, Sinai). *Limnol. Oceanogr.*, **24**, 799–822.

KÄMPF C. & PFENNIG N. (1980) Capacity of Chromatiaceae for chemotrophic growth. Specific respiration rates of *Thiocystic violacea* and *Chromatium vinosum*. *Arch. Microbiol.*, **127**, 125–35.

KAISER P. (1966) Ecologie des bactéries photosynthétiques. *Rev. Ecol. Biol. Sol.*, **3**, 409–72.

KOBAYASHI M. & HAQUE M. Z. (1971) Contribution to nitrogen fixation and soil fertility by photosynthetic bacteria. *Plant and Soil*, Spec. vol. (s.n.) 443–456.

KONINGS W. N. & VELDKAMP H. (1980) Phenotypic responses to environmental change. *Proc. Int. Symp. Micr. Ecol.*, **2**, 161–91.

KRASIL'NIKOVA E. N., PEDAN L. V. & KONDRAT'EVA E. N. (1977) Growth of the purple sulfur bacteria in dark under anaerobic conditions. *Microbiology*, **45**, 503–7.

KRASIL'NIKOVA E. N., PETUSHKOVA YU. P. & KONDRAT'EVA E. N. (1976) Growth of the purple sulfur bacteria *Thiocapsa roseopersicina* in the dark under anaerobic conditions. *Microbiology*, **44**, 631–4.

KUZNETZOV S. I. & GORLENKO W. M. (1973) Limnologische und mikrobiologische Eigenschaften von Karstseen der A. S. R. Mari. *Arch. Hydrobiol.*, **71**, 475–86.

LAANBROEK H. J., SMIT A. J., KLEIN NULEND G. & VELDKAMP H. (1979) Competition for L-glutamate between specialized and versatile *Clostridium* species. *Arch. Microbiol.*, **120**, 61–6.

LARSEN H. (1953) On the microbiology and biochemistry of the photosynthetic green sulfur bacteria. *Kl. Nor. Vidensk. Selsk. Skr.*, 1953(1), 1–205.

LOVITT R. W. & WIMPENNY J. W. T. (1979) The gradostat: a tool for investigating microbial growth and interactions in solute gradients. *Soc. Gen. Microbiol. Quarterly.*, **6**, 80.

LOVITT R. W. & WIMPENNY J. W. T. (1981) The gradostat, a bidirectional compound chemostat and its applications in microbiological research. *J. gen. Microbiol.*, **127**, 261–8.

MADIGAN M. T. & BROCK T. D. (1975) Photosynthetic sulfide oxidation by *Chloroflexus aurantiacus*, a filamentous photosynthetic gliding bacterium. *J. Bact.*, **122**, 782–4.

MATELES R. I., RYU D. Y. & YASUDA T. (1965) Measurement of unsteady growth rates in microorganisms. *Nature, Lond.*, **208**, 263–5.

MATHERON R. & BAULAIGUE R. (1976) Bactéries fermentatives, sulfato-réductrices et phototrophes sulfureuses en cultures mixtes. *Arch. Microbiol.*, **109**, 319–20.

MATHERON R. & BAULAIGUE R. (1977) Influence de la pénétration de la lumière solaire sur le dévelopment des bactéries phototrophes sulfureuses dans les environments marins. *Can. J. Microbiol.*, **23**, 267–70.

MEERS J. L. (1973) Growth of bacteria in mixed cultures. *CRC Crit. Revs. Microbiol.*, **2**, 139–84.

NIEL C. B. VAN (1931) On the morphology and physiology of the purple and green sulphur bacteria. *Arch. Mikrobiol.*, **3,** 1–112.
NIEL C. B. VAN (1936) On the metabolism of the Thiorhodaceae. *Arch. Mikrobiol.*, **7,** 323–58.
NIEL C. B. VAN (1955) Natural selection in the microbial world. *J. gen. Microbiol.*, **13,** 201–17.
NORTHCOTE T. G. & HALSEY T. G. (1969) Seasonal changes in the limnology of some meromictic lakes in southern British Columbia. *J. Fish Res. Bd. Can.*, **26,** 1763–87.
OKUDA A., YAMAGUCHI M. & KAMATA S. (1957) Nitrogen fixing microorganisms in paddy soils. III. Distribution of non-sulfur purple bacteria in paddy soils. *Soil Sci. Plant Nutr.*, Tokyo **2,** 131–3.
OREN A. & SHILO M. (1979) Anaerobic heterotrophic dark metabolism in the cyanobacterium *Oscillatoria limnetica*: sulfur respiration and lactate fermentation. *Arch. Microbiol.*, **122,** 77–84.
PADAN E. (1979) Impact of facultatively anaerobic photoautotrophic metabolism on ecology of Cyanobacteria (blue-green algae). *Adv. Micr. Ecol.*, **3,** 1–48.
PARSONS T. R. & TAKAHASHI K. (1973) The primary formation of particulate materials. In *Biological Oceanographic Processes* (Ed. by T. R. PARSONS & K. TAKAHASHI), pp. 58–105. Pergamon Press, Oxford.
PFENNIG N. (1961) Eine vollsynthetische Nährlösung zur selektieven Anreicherung einiger Schwefelpurpurbakterien. *Naturwiss.*, **48,** 136.
PFENNIG N. (1962) Beobachtungen über das Schwärmen von *Chromatium okenii*. *Arch. Mikrobiol.*, **42,** 90–95.
PFENNIG N. (1965) Anreicherungskulturen für rote und grüne Schwefelbakterien. *Zentr. Bakteriol. Parasitenk.*, **Abt. I,** Suppl. I, 179–89.
PFENNIG N. (1967) Photosynthetic bacteria. *Ann. Rev. Microbiol.*, **21,** 285–324.
PFENNIG N. (1978) General physiology and ecology of photosynthetic bacteria. In *The Photosynthetic Bacteria* (Ed. by R. K. Clayton & W. R. Sistrom), pp. 3–18. Plenum Press, New York.
PFENNIG N. & BIEBL H. (1976) *Desulfuromonas acetoxidans*, gen. nov. and sp. nov., a new anaerobic sulfur-reducing acetate-oxidizing bacterium. *Arch. Microbiol.*, **110,** 3–12.
PFENNIG N. & LIPPERT D. T. (1966) Über das vitamin B_{12}-bedürfnis phototropher schwefelbakterien. *Arch. Mikrobiol.*, **55,** 245–56.
PFENNIG N. & SCHLEGEL H. G. (1960) Anreicherungskulturen von Schwefelpurpurbakterien. *Generalversammlungsheft Ber. deutsche Bot. Ges.*, **73.**
PHILLIPS O. M. (1973) The equilibrium and stability of simple marine biological systems. I. Primary nutrient consumers. *Am. Nat.*, **107,** 73–93.
PIERSON B. K. (1973) The characterization of gliding filamentous phototrophic bacteria. Ph.D. Thesis, University of Oregon, USA.
PIERSON B. K. & CASTENHOLZ R. W. (1974a) A phototrophic gliding filamentous bacterium of hot springs, *Chloroflexus aurantiacus*, gen. and sp. nov. *Arch. Microbiol.*, **100,** 5–24.
PIERSON B. K. & CASTENHOLZ R. W. (1974b) Studies of pigments and growth in *Chloroflexus aurantiacus*, a phototrophic filamentous bacterium. *Arch. Microbiol.*, **100,** 283–305.
PIRT S. J. (1975) *Principles of Microbe and Cell Cultivation.* pp. 274. Blackwell Scientific Publications, Oxford.
POTTS M. & WHITTON B. A. (1979) pH and Eh on Aldabra Atoll. *Hydrobiologia*, **67,** 99–105.
POWELL E. O. (1967) The growth rate of microorganisms as a function of substrate concentration. In *Microbial Physiology and Continuous Culture*, Proc. 3rd Int. Symp. (Ed. by E. O. Powell, C. G. T. Evans, R. E. Strange & D. W. Tempest), pp. 34–55. HMSO, Porton Down.
RAVEN J. A. & BEARDALL J. (1981) The intrinsic permeability of biological membranes to H^+: significance for the efficiency of low rates of energy transformation. *FEMS Microbiol. Lett.*, **10,** 1–5.
ROELOFSEN P. A. (1935) On photosynthesis of the Thiorhodaceae. Ph.D. Thesis, University of Utrecht, The Netherlands.

RUTTNER F. (1962) *Grundriss der Limnologie*. pp. 332. Walter de Gruyter, Berlin.
SADLER W. R. & STANIER R. Y. (1960) The function of acetate in photosynthesis by green bacteria. *Proc. natn. Acad. Sci. USA*, **46**, 1328–34.
SCHLEGEL H. G. & PFENNIG N. (1961) Die Anreicherungskultur einiger Schwefelpurpurbakterien. *Arch. Mikrobiol.*, **38**, 1–39.
SCHRAMMECK J. (1934) Untersuchung über die Phototaxis der Purpurbakterien. *Beitr. Biol. Pflanz.*, **22**, 315–80.
SCHULZ E. (1936) Das Farbstreifensandwatt und seine Fauna. *Kieler Meeresf.*, **1**, 359–78.
SCHULZ E. & MEYER H. (1939) Weitere Untersuchungen über das Farbstreifensandwatt. *Kieler Meeresf.*, **3**, 321–36.
SENIOR E., BULL A. T. & SLATER J. H. (1976) Enzyme evolution in a microbial community growing on the herbicide Delapon. *Nature, Lond.*, **263**, 476–9.
SIEFERT E., IRGENS R. L. & PFENNIG N. (1978) Phototrophic purple and green bacteria in a sewage treatment plant. *Appl. Environm. Microbiol.*, **35**, 38–44.
SIREVÅG R. (1975) Photoassimilation of acetate and metabolism of carbohydrate in *Chlorobium thiosulfatophilum*. *Arch. Microbiol.*, **104**, 105–11.
SIREVÅG R. & ORMEROD J. G. (1977) Synthesis, storage and degradation of polyglucose in *Chlorobium thiosulfatophilum*. *Arch. Microbiol.*, **111**, 239–44.
SOROKIN YU. I. (1970) Interrelations between sulphur and carbon turnover in meromictic lakes. *Arch. Hydrobiol.*, **66**, 391–446.
SOROKIN YU. I. & DONATO N. (1975) On the carbon and sulphur metabolism in the meromictic Lake Faro (Sicily). *Hydrobiologia*, **47**, 241–52.
STANIER R. Y., ADELBERG E. A. & INGRAHAM J. (1976) *The Microbial World*. pp. 871. Prentice Hall Inc., Englewood Cliffs.
STEPHANOPOULOS G. & FREDRICKSON A. G. (1979) Coexistence of photosynthetic microorganisms with growth rates depending on the spectral quality of light. *Bull. Math. Biol.*, **41**, 525–42.
STUIVER M. (1967) The sulfur cycle in lake waters during thermal stratification. *Geochim. Cosmochim. Acta*, **31**, 2151–67.
SUCKOW R. & SCHWARZ E. W. (1963) Redox conditions and precipitation of iron and copper in sulphureta. In *Marine Microbiology* (Ed. by C. H. Oppenheimer), pp. 187–93. C. C. Thomas, Springfield.
TAKAHASHI M. & ICHIMURA S. (1968) Vertical distribution and organic matter production of photosynthetic sulfur bacteria in Japanese lakes. *Limnol. Oceanogr.*, **31**, 644–55.
TAYLOR P. A. & LEB. WILLIAMS P. J. (1975) Theoretical studies on the coexistence of competing species under controlled flow conditions. *Can J. Microbiol.*, **21**, 90–98.
TEMPEST D. W. (1970) The continuous cultivation of microorganisms. I. Theory of the chemostat. *Meth. Microbiol.*, **2**, 259–77.
TRÜPER H. G. & GENOVESE S. (1968) Characterization of photosynthetic sulfur bacteria causing red water in Lake Faro (Messina, Sicily). *Limnol. Oceangr.*, **13**, 225–32.
TRÜPER H. G. & PFENNIG N. (1966) Sulphur metabolism in Thiorhodaceae. III. Storage and turnover of thiosulphate sulphur in *Thiocapsa floridana* and *Chromatium* species. *Antonie van Leeuwenhoek, J. Microbiol. Serol.*, **32**, 261–76.
TRÜPER H. G. & SCHLEGEL H. G. (1964) Sulphur metabolism in Thiorhodaceae. I. Quantitative measurements on growing cells of *Chromatium okenii*. *Antonie van Leeuwenhoek, J. Microbiol. Serol.*, **30**, 225–38.
VAITUZIS Z. & DOETSCH R. N. (1969) Motility tracks: techniques for quantitative study of bacterial movement. *Appl. Microbiol.*, **17**, 584–88.
VELDKAMP H. & JANNASCH H. W. (1972) Mixed culture studies with the chemostat. *J. Appl. Chem. Biotechnol.*, **22**, 105–23.
VELDKAMP H. (1976) *Continuous Culture in Microbial Physiology and Ecology*. pp. 68. Shildon Co., Meadowfield Press, Durham.

WATANABE I. & FURUSAKA C. (1980) Microbial ecology of flooded rice soils. *Adv. Microbial Ecol.*, **4**, 125–68.
WETZEL R. G. (1973) Productivity investigations of interconnected lakes. I. The eight lakes of the Oliver and Walters chains, northeastern Indiana. *Hydrobiol. Stud.*, **3**, 91–143.
WETZEL R. G. (1975) *Limnology.* pp. 743. W. B. Saunders Co., Philadelphia.
WIJBENGA D.-J. & VAN GEMERDEN H. (1981) The influence of acetate on the oxidation of sulfide by *Rhodopseudomonas capsulata. Arch. Microbiol.*, **129**, 115–18.
WIMPENNY, J. W. T. (1981) Spatial order in microbial ecosystems. *Biol. Revs.*, **56**, 295–342.
WINOGRADSKI S. N. (1888) *Beiträge zur Morphologie und Physiologie der Bakterien.* Vol. I. Zur Morphologie und Physiologie der Schwefelbakterien. pp. 120. Verlag A. Felix, Leipzig.
YOON H., KLINZING G. & BLANCH H. W. (1977) Competition for mixed substrates by microbial populations. *Biotechnol. Bioeng.*, **19**, 1193–210.
ZEVENBOOM W. (1980) Growth and nutrient uptake kinetics of *Oscillatoria agardhii.* Ph.D. thesis, University of Amsterdam, The Netherlands.

APPENDIX. CULTIVATION METHODS

This section contains practical details of media, inocula, methods for enrichment and isolation of phototrophic bacteria. No extensive explanations are offered. We simply aim at opening the possibility for more people to grow these attractive organisms, no matter whether they want to perform sophisticated experiments or just like the look of fancy coloured bottles in their (north) window. Unjustly, the purple and green sulphur bacteria are considered more difficult to cultivate than the purple non-sulphur bacteria. For this reason, more attention is paid to the preparation of media for the phototrophic sulphur bacteria.

Growth media

Basically, media are made up by adding the following categories of ingredients:

1. inorganic salts to serve macronutrient requirements;
2. micronutrients or trace elements;
3. organic growth factors;
4. electron donors and carbon sources if not already added;
5. sources of energy.

Some components of a medium cannot be autoclaved in the same solution as the other components; other components cannot be autoclaved at all, and still others decompose or form inhibitory substances in the presence of oxygen. For these reasons, recipes may sometimes look very complicated and to some extent this is true for the media described here. Basically, however, media for growing phototrophic bacteria are simple and easy to prepare when the necessary precautions are taken. The use of stock solutions is convenient, but not essential.

A medium for growing purple and green sulphur bacteria

Final composition

NH$_4$Cl	300 mg
K$_2$HPO$_4$	300 mg
CaCl$_2$.2H$_2$O	200 mg
[a] MgCl$_2$.6H$_2$O or MgSO$_4$.7H$_2$O	200 mg/250 mg
KCl	200 mg
[b] NaCl	30 g
[c] Trace elements solution	10 ml
[d] Na$_2$CO$_3$ or NaHCO$_3$	2100 mg/1700 mg (ca. 20 mM)
Na$_2$S.9H$_2$O	100–700 mg (0.4–3 mM)
[e] Na$_2$S$_2$O$_3$	0–500 mg (0–3 mM)
[f] Na-acetate	0–250 mg (0–3 mM)
[g] Vitamin B$_{12}$	20 µg
[h] pH	6.8–7.5
distilled/demineralized water	1000 ml

[a] sulphate is produced during growth by sulphide oxidation. Its presence at the start may stimulate growth of sulphate-reducing bacteria, especially in combination with certain organic compounds, see [f].
[b] marine strains only.
[c] Pfennig & Lippert, 1966, see below.
[d] bicarbonate is usually recommended, but the carbonate can be autoclaved without loss of carbon dioxide.
[e] to augment growth, autoclave separately in neutral solution.
[f] may be replaced by other organic compounds, see [a].
[g] not required by all forms.
[h] adjust with HCl.

Preparation

Stock solution A

NH$_4$Cl	30 g
K$_2$HPO$_4$	30 g
pH	3 (HCl)
Water up to 1000 ml	
Store in refrigerator	

Stock solution B

CaCl$_2$.2H$_2$O	20 g
MgCl$_2$.6H$_2$O or MgSO$_4$.7H$_2$O	20 g/25 g
KCl	20 g
Water	up to 1000 ml
Store in refrigerator	

Stock solution C (Pfennig & Lippert, 1966)

Na_2-EDTA	500 mg
[a] $FeSO_4.7H_2O$	200 mg
[a] $ZnSO_4.7H_2O$	10 mg
$MnCl_2.4H_2O$	3 mg
H_3BO_3	30 mg
$CoCl_2.6H_2O$	20 mg
$CuCl_2.2H_2O$	1 mg
$NiCl_2.6H_2O$	2 mg
$Na_2MoO_4.2H_2O$	3 mg
pH	3 (HCl)
Water	up to 1000 ml

[a] can be replaced by chloride when desired.

It is convenient to make up larger quantities (e.g. 2 litre). To prevent growth (!), the solution can be made up more concentrated, or prepared at an even lower pH. Alternatively, the solution can be autoclaved in, e.g. 100 ml quantities. Once opened store in refrigerator.

Stock solution D

Cyanocobalamin (vitamin B_{12})	20 mg
Water	1000 ml
Store in refrigerator	

Stock solution E

Na_2CO_3 anh	20 g
Water	1000 ml

Autoclave in convenient volumes in screw-capped bottles. Do not use cotton-plugged containers (O_2!). Bottles, once opened, should not be re-used.

Stock solution F

Na_2S $7-9H_2O$	24 g (100 mM)
Na_2CO_3 anh	21 g (200 mM)
Water	1000 ml

Have autoclave ready to use. Dissolve salts in boiling (10 min!) water, dispense in convenient volumes, e.g. 10 ml in screw-capped tubes of 15 ml capacity; use dispenser, be careful with mouth-pipetting, sulphide is very toxic, and autoclave without delay. Tubes, once opened, should not be saved. This solution is also used for feeding (see below). The carbonate functions not only as carbon source but also as buffer for pH. In order to prevent a reduction of the buffering capacity due to growth upon repeated feeding, a corresponding amount of carbonate is added with each feed.

Stock solution G	
Na-acetate (anh)	8.2 g (100 mM)
Water	1000 ml

Autoclave in convenient volumes in screw-capped containers.

Stock solution H	
HCl	ca N

Autoclave in convenient volumes in screw-capped containers.

In order to prepare a litre of medium, add 10 ml each of the stock solutions A, B and C, and 1 ml of stock solution D to about 800 ml of distilled or demineralized water. The pH should be about 4 to prevent precipitation during autoclaving. Place stirring bar in bottle. Autoclave, allow to cool to room temperature. During the following operation, try to avoid the entrance of oxygen as much as possible: either have coffee before or after, but not during.

To the sterile mineral salt solution, add 100 ml of solution E: a fluffy whitish precipitate is formed which dissolves upon acidification. Add 4–30 ml of solution F: a slight red colouration will develop. Add solution H until desired pH is reached (about 20 ml is required). At a neutral pH the medium should be crystal clear and colourless. If yellow, repeat with a new batch, as this usually indicates the entrance of too much O_2 somewhere in the procedure; the medium is not entirely worthless, but long lag times will be observed. Fill bottle completely with sterile distilled water, leaving a pea-sized gas bubble to meet pressure changes. Store in the dark for at least one hour (or longer). A black precipitate (FeS) may be formed which disappears during growth.

To inoculate, replace a not too small volume (e.g. 5%) by the inoculum. Make sure a small bubble is present since completely filled bottles easily crack when the temperature rises. Again, store in the dark for one hour before placing the bottle in the light.

The whole procedure can be modified in such a way that bottles of, e.g. 100 or 150 ml are made up. Alternatively, such bottles can befilled with complete medium.

Inoculum

Many, if not all, small stagnant (anaerobic!) bodies of water such as ditches, forest ponds etc., especially when eutrophic, as can easily be judged from the presence of *Lemna*, contain representatives of the Chromatiaceae and the Chlorobiaceae. Often when organic substrates are freely available, the whole body of water may turn red as a consequence of the intense sulphate reduction. Such samples can be placed directly in the culture medium. Anaerobic marine mud is also an excellent inoculum (use NaCl in medium). Alternatively, Winogradsky columns (see Section 8.3.4) can be made up first.

Illumination

Use incandescent light sources. The intensity of irradiation should be between 100 and 300 lux, depending on the organisms (see Table 8.2, p. 170). An intensity of 100 lux is at a distance of about 50 cm from a 60 W bulb. Beware of excessive temperature rises—the limit is about 35°C.

Feeding

Slowly add about 3 ml of stock solution H(HCl) to 10 ml of stock solution F (sulphide + carbonate). Avoid excessive gas formation. The solution turns slightly yellow and may become turbid. If the turbidity does not disappear, too much acid has been added. Add 1–3 ml feeding solution per 100 ml culture; the sulphide concentration is then 0.8–2.3 mM. The weak acid sulphide is oxidized to the strong acid sulphate during growth. The resulting acidification is compensated for by an alkalinization as a result of carbondioxide fixation. The pH in sulphide/CO_2 media does not show dramatic changes; nevertheless occasional checking is recommended. If desired, the carbonate concentration may be doubled.

Purification

Add 1–2% agar to the mineral salt solution before autoclaving. Prepare the medium as described, preferably with the addition of thiosulphate and acetate to augment growth. Prepare 5% ascorbic acid solution, neutralize to pH = 6, sterilize by passing through a 0.2 μm membrane filter and add 10 ml per litre medium to give a final concentration of 0.05%. This solution is readily oxidized by air and should be freshly made. Dispense the medium in quantities of about 10 ml in cotton-plugged test tubes. Add a small volume (about 1 ml) of the enrichment to the first tube, mix by inverting the tube (compromise between ideal mixing and avoidance of aeration). Add about 1 ml from the first tube to the second, and so on until 10 tubes are inoculated. When inoculated, the tubes are placed in a waterbath (10–15°C) to speed up solidification of the agar. Thereafter, a few ml of a 1/1 mixture of liquid and solid paraffin (sterilized by autoclaving) may be added to each tube to prevent the entrance of oxygen during incubation. Alternatively, inoculated tubes (larger size) may be emptied in sterile petri dishes which are placed immediately in transparent anaerobic incubators. Incubate in the light. Colonies become visible in 1–3 weeks. Pick a not too small colony (either cut the tube and gently blow the agar column into a sterile petri dish, or open the jar and remove a plate colony with a pasteur pipet) and place immediately in a small volume of complete medium. Then, check microscopically for the presence of the desired organism. Thereafter, the whole procedure can be repeated if necessary.

Discard tubes in which growth between glass and agar has occurred.

A medium for growing purple non-sulphur bacteria

Final composition

NH_4Cl	300 mg
[a] KH_2PO_4	300 mg
$CaCl_2.2H_2O$	50 mg
[b] $MgCl_2.6H_2O$ or $MgSO_4.7H_2O$	200 mg/250 mg
NaCl	400 mg
[c] trace elements solution	10 ml
[a] $Na_2CO_3/NaHCO_3$	2000 mg
[d] yeast extract	200–1000 mg
[e] vitamin solution	10 ml
[f] Na_2-succinate	1000 mg
[g] Na-ascorbate	500 mg
pH	6.7–7.3
distilled water	1000 ml

[a] alternatively, raise the phosphate concentration to 1000 mg and reduce the carbonate concentration to 500 mg
[b] the presence of sulphate may stimulate growth of sulphate-reducing bacteria. Sulphide is inhibitory to most Rhodospirillaceae.
[c] Pfennig & Lippert, 1966, see stock solution C.
[d] may be replaced by the vitamin solution
[e] Pfennig, 1965, see below
[f] may be replaced by, e.g. lactate (sterilize separately in 10% solution at neutral pH).
[g] can be omitted for most strains; if needed sterilize by membrane filtration (see purification).

Vitamin solution (Pfennig, 1965)

biotin	2 mg
nicotinamide	20 mg
p-aminobenzoic acid	10 mg
panthothenic acid	5 mg
pyridoxin	50 mg

Sterilize by filtration through 0.2 μm membrane filters.

The preparation of the medium is done in much the same way as described for the medium for Chromatiaceae and Chlorobiaceae. For enrichments, sterilization can often be omitted.

Suggested reading

DREWS G. (1965) *Zentr. Bakteriol. Parasitenk.*, Abt. I Suppl. I 170–8.
DSM *Catalogue of strains* (1977) Ges. Strahlen-und Umweltforsch. Munchen.
PFENNIG N. (1965) *Zentr. Bakteriol. Parasitenk.* Abt. I Suppl. I 179–189
VAN NIEL C. B. (1971) *Methods in Enzymology* XXIII 3–28.

Chapter 9. Genetics and Molecular Biology

B. L. MARRS

9.1 INTRODUCTION

The phototrophic bacteria exhibit a wide range of interesting biological phenomena, many of which have been treated in the preceding chapters of this book. The genetics and molecular biology of this group are equally fascinating areas, and furthermore they provide tools, techniques and insights that can aid in understanding most of the other areas under investigation in these organisms.

For the purposes of this chapter, molecular biology shall be construed as the study of processes involved in the maintenance and expression of genetic information. Only certain aspects of the molecular biology of phototrophic bacteria have been investigated in detail, because most such processes are expected to be similar to those already demonstrated and more easily pursued in *Escherichia coli* or elsewhere. On the other hand, the great metabolic and energetic flexibility of some of the organisms of this group, especially the rhodopseudomonads, affords special opportunities to examine regulatory mechanisms at the molecular level. Studies directed at probing those mechanisms have yielded most of our knowledge about the molecular biology of these species.

Genetics of the anoxygenic phototrophic bacteria is confined as of this writing to the genus *Rhodopseudomonas*, but for *Rp. capsulata* and *Rp. sphaeroides* workable systems for genetic analysis have been established. Two conjugative systems have been described for *Rp. sphaeroides* and one for *Rp. capsulata*. Each of these is based upon self-mobilizable, antibiotic resistance-conferring plasmids called R-factors, and each provides a means of mapping large blocks of chromosomal DNA. The only tool for mapping genetic fine structure is the gene transfer agent (GTA) of *Rp. capsulata*, and that organism is thus best suited for genetic studies, since both types of mapping are possible. In addition to classical genetics, *Rp. capsulata* has been used as both a source of DNA and a host in recombinant DNA experiments, and thus its utility in cloning experiments is established.

The molecular biology and genetics of these bacteria have been reviewed by

Marrs *et al.* (1977), Saunders (1978), Marrs (1978a) and Gray (1978). Kaplan (1978) has reviewed the control of photosynthetic membrane development. In the short time since those reviews were written there have been substantial advances in the genetics and molecular biology of the phototropic bacteria, and, this chapter focuses primarily on these recent developments.

9.2 PROTEIN SYNTHESIS

It has long been recognized that the phototrophic bacteria afford the potential of a system in which to study changes in protein synthesis associated with the differentiation of the photosynthetic membrane (Lascelles, 1959; Sistrom, 1962; Sistrom, 1963; Cohen-Bazire & Kunisawa, 1963; Gibson *et al.*, 1963; Higuchi *et al.*, 1965).

Many of these studies were stimulated by the germinal work of Cohen-Bazire, Sistrom and Stanier (1957) which quantified the inhibitory effects of light and oxygen on pigment content in *Rp. sphaeroides* membranes. A recent study by Madigan, Cox and Gest (1981) provides a lovely example of the significance to the organism of the ability to regulate the expression of the photosynthetic apparatus, and at the same time illustrates how the metabolic versatility of the rhodopseudomonads can aid in the study of such regulation. *Rp. capsulata* can be grown anaerobically in darkness if provided with fermentable substrates and an accessory oxidant (Yen & Marrs, 1977; Madigan & Gest, 1978). These conditions would be expected to promote extensive development of the photosynthetic apparatus, since the two known environmental signals for limiting its synthesis, light and O_2, are both absent. Indeed, for the first few subcultures in anaerobic dark conditions an increase in pigment content (relative to cell protein) is observed. Cells harvested from subsequent subcultures, however, show lower and lower pigment contents. After about 20 transfers the photopigments are barely detectable. When individual cells are cloned from such subcultures, a variety of mutants blocked in the synthesis of the photosynthetic apparatus are obtained. Similar observations were made by Uffen *et al.* (1971) on *Rhodospirillum rubrum* cultured anaerobically in the dark. Evidently the burden of synthesizing the unused photosynthetic apparatus is so great that mutants that have lost the ability to make it have a strong selective advantage. Perhaps this observation provides a rationalization for why most rhodopseudomonads grow only slowly or not at all under standard fermentative conditions, even though they appear to have all the enzymes needed for fermentative growth (Madigan *et al.*, 1980). *Rp. capsulata* has not evolved a mechanism for repression of the photosynthetic apparatus under fermentative conditions. The evolutionary advantages of retaining the photosynthetic apparatus through periods of darkness might outweigh the ability to multiply more rapidly in the dark. In any event it is clear that mutational loss of the inappropriate synthesis of the apparatus gives an organism a marked growth advantage during fermentation, and a physiological mechanism for repression during respiratory growth avoids the mutational loss of photosynthetic ability.

9.2.1 Messenger RNA

Early studies established that concomitant protein synthesis was essential for pigment synthesis (Bull & Lascelles, 1963; Sistrom, 1963), but attempts to demonstrate changes in RNA metabolism associated with increased photosynthetic membrane synthesis were negative or indecisive (Lessie, 1965b; Gray, 1967; Cost & Gray, 1967; Yamashita & Kamen, 1968; Witkin & Gibson, 1972a and b; Chow, 1976a). Differences in the proportion of RNA that was message and in the half-life of message were noted during transitions affecting the amount of photosynthetic membrane synthesized, but they could not be unambiguously related to the regulation of membrane formation. Although some of these authors chose to interpret their inability to demonstrate the appearance of a new mRNA class upon membrane induction to be suggestive of translational control, it seems more likely that the negative findings were a result of methods that were not sensitive enough to detect the expected transcriptional differences. If transcriptional controls are involved in the regulation of photosynthetic membrane synthesis, their existence could be demonstrated by hybridization experiments with specific DNA probes. Appropriate probes have recently become available (see Section 9.5.2), so this question should soon be settled.

Chow (1977; 1978) has reported the occurrence of three forms of DNA-dependent RNA polymerase in heterotrophically grown *Rs. rubrum*. None of the activities reported resembles a typical bacterial DNA-dependent RNA polymerase, but instead they have unusual subunit compositions, low divalent cation requirements and are resistant to rifampicin and streptovaricin. Chow emphasizes the similarities between these properties and eukaryotic RNA polymerases. Since Chow himself (1976a) has shown that rifampicin causes a rapid breakdown of pulse-labelled RNA in *Rs. rubrum*, rifampicin presumably inhibits the mRNA-producing RNA polymerase. It therefore seems likely that the mRNA-synthesizing RNA polymerase was not among those studied *in vitro*. The enzymes studied showed polyadenylic acid polymerase activity, and it therefore seems possible that they were related to polynucleotide phosphorylase, which can polymerize RNA and demonstrates a polynucleotide primer requirement upon purification (Ingram, 1972).

The existence in *Rs. rubrum* of polyadenylated mRNA coding for reaction centre and light-harvesting polypeptides has been reported (Majumdar & Vipparti, 1980). Since polyadenylated mRNA is characteristic of eukaryotes and has never before been reported in a prokaryote, it would be interesting if this report were correct. Before this finding can be accepted, however, it must be demonstrated that the mRNAs in question, which were isolated by binding to an oligo(dT)-cellulose column, actually contain 3′ poly A tails, as opposed to internal A-rich sequences. Furthermore, while the results do show translation products that co-migrate with some polypeptides of the photosynthetic apparatus, they do not demonstrate that the oligo(dT)-cellulose-binding mRNA fraction is enriched in reaction centre or light-harvesting-specific mRNAs. This is because co-migration in one SDS-PAGE system is not a sufficient criterion for identity. Furthermore, the translation products of the rest of the mRNA (not bound to oligo(dT)-cellulose) were not examined, and thus conclusions about enrichment are not possible. Specific DNA probes for the reaction centre and light harvesting genes would facilitate the clarification of these observations.

9.2.2 Transfer RNA

The tRNA composition of *Rp. sphaeroides* has been examined by DeJesus and Gray (1971), Shepherd and Kaplan (1975) and Razel and Gray (1978). The object of these studies was to search for isoaccepting species of tRNA that varied with the expression of the photosynthetic apparatus, since such compositional differences could signify an underlying regulatory mechanism. Multiple isoaccepting species of tRNA were found, but for most amino acids these were the same for phototrophic as for dark-grown cultures. Both laboratories found, however, that phenylalanyl and tryptophanyl tRNA's exhibited marked differences in composition in the two cell types. There are two tryptophanyl tRNA's that may be separated by benzoylated DEAE chromatography in both aerobic and phototrophic cultures, but the ratio of major to minor species is 2.0 in tRNA from aerobic cells and 8.0 from phototrophic. The different ratios were shown to reflect the *in vivo* situation, and not to be an artifact of *in vitro* charging.

Four species of $tRNA^{Phe}$ were identified. In phototrophic cultures more than 80% of the tRNA that could be charged with phenylalanine was $tRNA^{Phe}$ III. In cultures grown aerobically for more than six divisions $tRNA^{Phe}$ II represents about 80% of the total, and $tRNA^{Phe}$ III only 8%. From studies on the effects of chloramphenicol and rifampin on the transition from one pattern of tRNA's to the other, Razel and Gray (1978) suggested that $tRNA^{Phe}$ II is modified to give rise to $tRNA^{Phe}$ III. The modifying enzyme is synthesized during phototrophic but not aerobic growth. The enzyme, if present from prior growth, is proposed to be active under either growth condition. Bacteriochlorophyll synthesis is required for synthesis of the photosynthetic organelles of *Rp. sphaeroides* and *Rp. capsulata* (Kaplan, 1978). It would be interesting to see whether mutants blocked in bacteriochlorophyll synthesis would undergo the shift in tRNA patterns when transferred from high to low aeration. This would position the timing of the tRNA effect with respect to the block in membrane synthesis caused by a bacteriochlorophyll defect.

9.2.3 Ribosomes

The ribosomes of phototrophic bacteria are similar to those of most other bacteria. Those examined have sedimentation constants near 66S, and they dissociate into subunits of 29S and 45S (*Rs. rubrum*, *Rp. sphaeroides*, *Rp. palustris* and *Rp. gelatinosa*; Gray, 1978). The larger ribosomal subunit from most species of phototrophic bacteria contains RNA molecules of mol. wt. about 1.1×10^6 and 3.5×10^4, and the small subunit one of about 0.45×10^6. Three *Rhodopseudomonas* spp. share a rare trait in that they appear to contain no 1×10^6 rRNA (Lessie, 1965a; Marrs & Kaplan, 1970; Robinson & Sykes, 1971; Gray, 1978). It has been demonstrated that in *Rp. sphaeroides* a 1.1×10^6 mol. wt. rRNA precursor is formed, but it is cleaved during the maturation of the 45S subunit to give the 0.53 and 0.42×10^6 mol. wt. fragments that are found on the mature large subunit (Marrs & Kaplan, 1970). The pathways of RNA processing and methylation in this species have been described (Gray, 1973; Gray, 1978). This rare type of rRNA processing is also observed in *Paracoccus denitrificans*, a Gram-negative aerobe. Furthermore, the 16 and 23S rRNAs of *Paracoccus* show considerable sequence homology with the rRNAs of *Rp. capsulata* and *Rp. sphaeroides* (MacKay *et al.*, 1979; Gibson *et al.*, 1979). These observations, together with remarkable similarities in Cytochrome c_2 sequences have led to the proposal that *Paracoccus* may have evolved from a rhodopseudomonad by loss of photosynthetic ability (Dickerson *et al.*, 1976; Fox *et al.*, 1980). Since *Paracoccus* has been singled out as that organism most like the mitochondrion of eukaryotic cells (John & Whatley, 1975), the rhodopseudomonads take on added significance in our anthropocentric world. As Dickerson (1980) put it: 'Human beings are the metabolic offspring of defective purple photosynthetic bacteria'.

There have been a few studies that searched for differences between ribosomes from heterotrophic and phototrophic cells. Mansour & Stachow (1975) isolated ribosomal subunits from phototropic and aerobic cultures of *Rp. palustris* and compared their protein compositions. Previous studies had shown that the 29S and 46S ribosome subunits from phototrophic cells comprise 23 and 28 different proteins respectively (Bhatnagar & Stachow, 1972). Mansour and Stachow (1975) reported that there were four differences between the proteins of the 29S ribosomal subunits of cell from the two types of cultures. The differences involved three proteins that appeared to have altered electrophoretic mobility and one protein that was present in aerobic ribosomes but absent in phototrophic. The authors interpreted their data to suggest that *Rp. palustris* uses two distinct populations of ribosomes for phototrophic and aerobic growth. A peculiarity of this study was the means by which the aerobic cells were obtained. A fully grown phototrophic culture was incubated aerobically with shaking in darkness. A 100 hour transition period ensued during which there was no growth and ribosome degradation occurred. Eventually growth resumed, and the culture was harvested as a source of aerobic cells. A long lag period is not typical of a phototrophic–aerobic transition for most purple non-sulphur bacteria, and phototrophic cultures usually achieve a cell density that is much greater than any obtainable by aerobic growth (Marrs, unpublished observations). One wonders if the culture that eventually grew out consisted of the survivors of a period of starvation. If so, the results must be interpreted with caution.

Chow (1976b and c) has prepared cell-free protein synthesizing systems from phototrophic and aerobic Rs. rubrum. He found that only minor differences in protein synthesizing ability exist between these two systems; however, he did report that ribosomes from aerobic cells were less stable than those prepared from phototrophic cells. An inhibitor and an activator of protein synthesis are reported to be present in extracts of aerobic cells. The inhibitor is dialysable and inhibits protein-synthesis non-specifically, while the activator is a non-dialysable RNA molecule, which stimulates only phototrophic ribosomes of Rs. rubrum. Qualitative differences between the phototrophic and aerobic ribosomal proteins were described, but these seemed to depend upon the gel system used, and it is hard to assess the contributions of contaminating proteins in these studies.

The recent reports about the process of protein synthesis in phototrophic bacteria can be summarized as provocative but wanting substantiation. Several authors have interpreted their data to suggest that these bacteria possess a more eukaryotic-like protein synthetic pattern than do other prokaryotes. If true, this would be of obvious interest.

9.3 PLASMIDS

Plasmids are extra-chromosomal, genetic elements that are widespread in the prokaryotic world. Plasmids are usually covalently closed circular DNA molecules which can replicate autonomously, and they are known to code for a wide range of phenotypic traits including antibiotic resistance, mobilization of DNA, tumor induction, root nodulation, nitrogen fixation, pathogenicity, and the metabolism of unusual natural products (Nuti et al., 1979; Fennewald et al., 1978). These traits are generally dispensable. The significance of plasmids in relation to phototrophic bacteria is twofold. Of primary importance is the ability of certain plasmids to mobilize chromosomes and thus provide genetic exchange mechanisms. This section discusses the expression of exogenous-plasmid-borne genes and the mobilization and stability of the plasmids themselves, and chromosome mobilization is discussed in Section 9.4. The fact that many phototrophic bacteria possess indigenous cryptic plasmids of unknown function has intrigued investigators for many years. Could the genes for photosynthesis be plasmid-borne? The evidence seems to be mounting against this idea, but no final conclusion can be drawn at this time.

9.3.1 Indigenous plasmids

Indigenous extrachromosomal DNA has been reported for *Rp. sphaeroides* (Suyama & Gibson, 1966; Gibson & Niederman, 1970; Saunders *et al.*, 1976), *Rp. capsulata* (Hu & Marrs, 1979), *Rs. rubrum* (Kuhl & Yoch, 1981) and *Chromatium* D (Suyama & Gibson, 1966). No function is known for any of these plasmids. A naturally occurring viral R-plasmid has been described for *Rp. sphaeroides* strain RS601 (Pemberton & Tucker, 1977; Tucker & Pemberton, 1978). This unusual entity codes for ampicillin resistance and bacteriophage production and exists as covalently closed circular, supercoiled DNA in both the Rϕ6P phage particle and the bacterial cytoplasm.

Most strains of both *Rp. sphaeroides* and *Rp. capsulata* contain multiple species of plasmids. The former species contains two plasmids with the same buoyant density, 1.717 g/cm^3, and molecular weights of 28 and 66×10^6 atomic mass unit (awu), and one plasmid with a buoyant density of 1.724 g/cm^3 and a molecular weight of 75×10^6 awu (Gibson & Niederman, 1970; Saunders *et al.*, 1976). Saunders *et al.* (1976) described a non-phototrophic mutant of *Rp. sphaeroides* which contained a 34×10^6 awu plasmid in place of the 28×10^6 one, but it was not possible to establish a causal relationship between the altered plasmid and the altered photosynthetic ability. *Rp. capsulata* strain BH9 contains plasmid molecules of 75 and 94×10^6 awu (Hu & Marrs, 1979). Determination of reassociation kinetics (C_0t analysis) of unfractionated *Rp. capsulata* plasmids indicated that approximately 1.6×10^8 awu of unique sequence DNA was present. C_0t analysis in the presence of added whole cell DNA showed that approximately 10% of the total BH9 DNA could hybridize to plasmid sequences. Thus there are about 2.8×10^8 awu of plasmid sequences per genome equivalent of BH9 DNA, and one copy of each of the known plasmids could account for about 1.7×10^8 awu. The excess mass estimated by C_0t analysis could signify that two copies of one of the larger plasmids were usually present per genome equivalent.

The plasmid found in *Rs. rubrum* has a mass of 36×10^6 awu. Phototrophically incompetent mutants isolated following ethidium bromide treatment were found to have 'lost' the plasmid, but revertants 'regained' it (Kuhl & Yoch, 1981, and personal communication). This suggests that perhaps the plasmid is an episome, capable of reversible integration into the chromosome. The loss of photosynthetic ability in the integrated state could be due to the inactivation of a chromosomal gene upon integration, or to the loss of expression of plasmid genes upon integration. Further studies with *Rs. rubrum* should prove interesting.

Two lines of evidence from *Rp. capsulata* tend to indicate that the genes for photosynthesis are not normally plasmid-borne in that species, but a logically tight conclusion cannot yet be drawn. Conjugative studies (Marrs, 1981) have shown that the genes for photosynthesis are genetically linked to genes for tryptophan synthetase and rifampicin resistance. Since these latter genes are typical chromosomal markers, the genes for photosynthesis must be chromosomal at least part of the time. An episomal linkage group for photosynthesis would be consistent with these observations. (Since episomes can integrate into chromosomes, they can establish genetic linkage between markers in the two replicons.) Similar observations have been made in *Rp. sphaeroides* (Pemberton & Bowen, 1981). The second piece of experimental evidence is the result of having available cloned fragments in the photopigment region. One such fragment was used in a Southern transfer and hybridization protocol (Southern, 1975) to probe for homologies among restriction endonuclease digests of chromosomal and plasmid DNAs prepared by dye-buoyant density gradient centrifugation. The chromosomal DNA fraction showed hybridization of the probe to a band of DNA of the appropriate size, whereas only a very faint band of hybridization was observed in the plasmid DNA preparation. The faint hybridization could represent a slight chromosomal DNA contamination of the plasmid band, or a partial homology between some plasmid sequence and the probe. However, since there is precedence for the existence of 'megaplasmids', plasmids of about 200×10^6 awu which code for nodulation, hydrogen uptake and nitrogen fixation in *Rhizobium* species (Brewin *et al.*, 1981), and since megaplasmids might be exceptionally sensitive to single strand nicking and therefore band with the chromosomal DNA in a dye-buoyant density gradient, the possibility of a megaplasmid location for photosynthesis genes remains a logical possibility. Megaplasmids have not been reported in the phototrophic bacteria.

9.3.2 Exogenous plasmids

Olsen and Shipley (1973) demonstrated that the plasmid R1822 could be transferred by conjugation from *Pseudomonas aeruginosa* to *Rp. sphaeroides* and *Rs. rubrum*. R1822 specifies resistance to carbenicillin (or ampicillin), tetracycline and kanamycin (or neomycin), and belongs to the P incompatibility group of R factors, characterized by their wide host-range among Gram-negative bacteria. It has subsequently been demonstrated that R1822, now called RP1, is identical with plasmids RP4, R68 and RK2 (Burkardt *et al.*, 1979), so studies conducted with any of these plasmids are directly comparable. Olsen and Shipley reported that carbenicillin resistance was not stably maintained in either *Rs. rubrum* or *Rp. sphaeroides* upon cloning on non-selective media, and they inferred that the R1822 plasmid was not stable in those hosts.

Sistrom (1977) demonstrated that R68.45 could be transferred from *P. aeruginosa* to both *Rp. sphaeroides* and *Rp. gelatinosa*. R68.45 is a derivative of R68 with enhanced chromosome mobilizing ability (Hass & Halloway, 1976). Sistrom was able to demonstrate chromosome transfer in *Rp. sphaeroides* mediated by R68.45, thus establishing the first conjugational system for any phototrophic bacterium. This genetic exchange system is discussed in Section 9.4.2. Sistrom showed that both neomycin and carbenicillin resistance were expressed in *Rp. gelatinosa*, but only neomycin resistance was expressed in *Rp. sphaeroides*. Since *Rp. sphaeroides* strains bearing R68.45 could transfer both neomycin and carbenicillin resistance to *Rp. gelatinosa*, Sistrom concluded that the penicillinase gene was present but not expressed in the former species.

Miller and Kaplan (1978) mobilized RP4 from *E. coli* into *Rp. sphaeroides* and found that the plasmid was stably maintained; however, carbenicillin resistance was not expressed after the initial transfer. The genetic determinant for carbenicillin resistance was not lost, as was demonstrated by the transfer of RP4 back into *E. coli* using tetracycline resistance as a selective marker. All tetracycline resistant clones of *E. coli* were also carbenicillin resistant. The kanamycin and tetracycline resistance markers of RP4 were expressed in *Rp. sphaeroides*, but the tetracycline resistance marker was not stably maintained in the absence of tetracycline. After repeated subculture in the presence of tetracycline a stable variant was obtained (Miller & Kaplan, 1978). They also reported that RP4 and related plasmids could be established in *Rp. capsulata*. Jasper *et al.* (1978) also observed the transfer of RP1 from *E. coli* to *Rp. capsulata* and the expression of kanamycin and tetracycline but not ampicillin resistance.

Tucker and Pemberton (1979a; b) tested the transmissibility of a number of plasmids from *E. coli* into *Rp. sphaeroides*. In addition to R68.45 and RP4, they found that P incompatibility group plasmids R751 and R702 could be transferred and maintained in *Rp. sphaeroides* as could the W group plasmids R388 and S-a. Transfer was not observed for plasmids RI and RI-16 of incompatibility group FII, R40a and R57b of group C, or R64 of group I. Markers carried by the P and W group plasmids were expressed in *Rp. sphaeroides*, with the exception of carbenicillin resistance and sensitivity to the male specific phages PR11 and PRR1. Surface exclusion and incompatibility properties were said to have been observed. A low frequency of chromosome transfer (10^{-7} to 10^{-9} per donor cell) was reported to be promoted by each of these plasmids. Tucker and Pemberton (1979b) also reported on the transfer of RP4::Mu *cts* 62 into *Rp. sphaeroides*. Although the thermosensitive nature of the Mu cointegrate was not expressed, Mu phage were released. Spontaneous curing of RP4::Mu *cts* 62 was frequently observed, leading to

Mu-negative, kanamycin-sensitive segregants. These experiments open up the possibility of *in vivo* genetic engineering in *Rp. sphaeroides* (Dénarié *et al.*, 1977). Yu *et al.* (1981) have performed similar experiments in *Rp. capsulata*. They transferred RP1, R68.45 and RP4::Mu *cts* 61 into *Rp. capsulata* strain 37b4 from *E. coli*, and observed the expression of kanamycin, tetracycline and ampicillin resistance for each plasmid-bearing strain. This is in contrast to the results of Jasper *et al.* (1978) who reported that ampicillin resistance was not expressed in *Rp. capsulata* strain SB1003 carrying RP1. Perhaps differences between strains SB1003 and 37b4 affect the expression of the penicillinase gene on RP1, which is known to be carried as the transposon, Tn*1*. Yu *et al.* (1981) also reported the production of Mu phages from *Rp. capsulata* carrying RP4::Mu *cts* 61, although others have not observed this production (Wall, personal communication). A low level chromosomal mobilization is mediated by RP1 and its derivatives pLM2, R68.45 and RP4::Mu *cts* 61 in *Rp. capsulata* (Yu *et al.*, 1981; Marrs, 1981).

Pemberton and Bowen (1981) have transferred RP1::Tn*501* into *Rp. sphaeroides*, and this plasmid promotes high frequency chromosome transfer. Tn*501* codes for mercury resistance, but the authors did not indicate whether this property is expressed in this host.

Jasper *et al.* (1978) transferred the non-self-transmissible plasmid colE1::Tn*5* into *Rp. capsulata* by transformation. This plasmid was stably maintained and the kanamycin resistance encoded by Tn*5* was expressed.

Most of the preceding experiments concerned with the introduction of exogenous plasmids into phototrophic bacteria have been motivated by the need to develop conjugative genetics and cloning capabilities in these bacteria. The next section deals directly with those topics.

9.4 GENETIC TRANSFER

Classical genetic analysis, the study of the arrangement of genes, mutations and regulatory elements by recombination and complementation, requires mechanisms for the transfer of DNA from one organism to another. The nature of each transfer mechanism, especially the size of the DNA fragments transferred, determines in large part the range and resolving power it affords. The rate of discovery of genetic transfer mechanisms for phototrophic bacteria has been relatively slow, and this, rather than a lack of need or utility, has retarded the development of the biochemical genetics approach to understanding these bacteria. It is now theoretically possible to perform genetic analyses in the absence of genetic transfer by cloning, sequencing and translating DNA *in vitro*, but classical genetics is still the more practical first approach to most problems. Fortunately, the discovery and development of genetic transfer systems has accelerated in recent years, and for one species, *Rp. capsulata*, all the tools necessary for classical genetic analysis are at hand.

9.4.1 Capsduction

The first genetic transfer mechanism discovered for a phototrophic bacterium was capsduction in *Rp. capsulata* (Marrs, 1974). Capsduction resembles transduction in that phage-like particles are the vectors that carry DNA from donor to recipient cells, yet the process is fundamentally different from transduction in that no virus is associated with particle production.

The particles that serve as vectors for capsduction are called gene transfer agents or GTA. Electron micrographs of GTA reveal particles with the overall morphology of a tailed bacteriophage. The head is 30 nm in diameter and appears icosahedral with short apical spikes. The tail is of variable length and is joined to the head by a collar. The tail is 5–6 nm in diameter, cross striated at 3–4 nm intervals, and ends in tail fibres. DNA extracted from GTA is in the form of linear, double-stranded molecules about 3×10^6 awu in molecular weight. The DNA has the same complexity, determined by $C_0 t$ analysis, as the genomic DNA of *Rp. capsulata*. Digestion of GTA DNA with the restriction endonuclease *Hpa*II, which recognizes a four base target sequence, shows no sequence that gives rise to a distinct band upon gel electrophoresis. Both the $C_0 t$ analysis and the restriction endonuclease results show that within the limits of resolution no virus-like DNA is present in GTA particles, and all the DNA of the genome is packaged uniformly, i.e., with no one sequence predominating (Yen *et al.*, 1979). Genetic and physical evidence indicates that DNAs from exogenous and indigenous plasmids alike are packaged in GTA particles (Jasper *et al.*, 1978; Hu & Marrs, 1979). Biological assays as well as the above-mentioned physical tests, indicate that there is no viral activity associated with GTA production. Searches have revealed no plaque-forming or killing activity in active, purified GTA preparations, or in crude preparations from most strains. GTA preparations are not infectious, since strains of *Rp. capsulata* that do not naturally produce GTA do not begin to produce it after GTA exposure, even though they may be quite capable of receiving genetic information via GTA particles (Wall *et al.*, 1975; Marrs, 1978a; Yen *et al.*, 1979).

A comment on the discovery of capsduction seems worthwhile. This unusual genetic exchange mechanism was recognized because the method used in the search for genetic exchange was very general and did not preclude unexpected mechanisms (Marrs, 1974). Fresh isolates of Rhodospirillaceae were tested for genetic exchange processes by mixing pairs of cultures and testing for recombinants after a period of mixed growth. A rifamipicin- and a streptomycin-resistant strain of each isolate were first selected, and then the progeny of mixed cultures were challenged with both antibiotics simultaneously. Cultures that showed rifampicin-streptomycin double resistance mutants in excess of unmixed control cultures were examined for the mechanism of genetic exchange, and capsduction was found. Similar phenomena might exist in other bacteria but go unnoticed, because in screening for genetic exchange systems one tends to look for mechanisms that are already known.

The distribution of gene transfer activity among isolates of *Rp. capsulata* is widespread. In one survey of 33 wild type strains, 19 were found to be capable of producing GTA, and 25 showed competence as recipients (Wall *et al.*, 1975). The GTA produced by 11 strains were compared immunologically, and a high degree of cross reactivity was found, indicating that similar particles were responsible for the gene transfer activity in each case (Yen *et al.*, 1979). No evidence for GTA activity was found among the other species of phototrophic bacteria tested, which included eight strains of *Rp. palustris*, three of *Rp. sphaeroides*, four of *Rp. gelatinosa*, one of *Rp. viridis*, five of *Rs. rubrum*, one of *Rs. molischianum*, one of *Rs. photometricum*, one of *Rs. tenue*, and one strain of *Rhodomicrobium vannielii* (Wall *et al.*, 1975).

The genetic characteristics of capsduction are amenable to mapping and strain construction. The DNA fragments introduced by GTA are not capable of autogenous replication, so recombination between the incoming fragment and the resident chromosome must occur for a stable genetic event to occur. The 3×10^6 awu fragment can introduce about five genes at a time, and from this fragment a piece may be integrated. Since the amount of DNA transferred is small, strains constructed via GTA differ only minimally from the recipient genotype. The map function that describes the relationship between the frequency of two points being co-transferred (ϕ) and the distance between the two points (d) is: $\phi = (1-d)^2$ (Yen & Marrs, 1976). This is consistent with a model in which fixed lengths of DNA with randomly distributed starting points are packaged in GTA particles and transferred to the recipient. A single crossover or breakage and refusion then occurs at any point of homology between the fragment and the chromosome, and donor DNA from that point to one terminus of the fragment then replaces the corresponding resident DNA, which is subsequently lost. Several genetic transfers can occur independently in one recipient cell if several donor GTA particles are taken up. Capsduction does not result in any detectable immunity to subsequent GTA uptake, and many strains are self fertile, so strains may be constructed one step at a time by a series of GTA-mediated crosses (Marrs, 1978b).

This model of capsduction is supported by recent restriction endonuclease mapping of the photopigment region. In the map function derived from the model, d = distance relative to the length of the DNA carried in the GTA, i.e., 1 map unit equals the amount of DNA packaged in a single particle. The physical estimate of GTA DNA molecular weight by gel electrophoresis was about 2.8×10^6 awu, hence that should be equivalent to one map unit. As will be seen below, the physical size of 1 map unit estimated by restriction endonuclease mapping is 3.0×10^6 awu, an excellent agreement.

The genetic map of the photopigment region (Fig. 9.1) constructed by capsduction is remarkably additive and internally consistent, and parts of it have been confirmed by restriction endonuclease mapping. The *crt* genes are

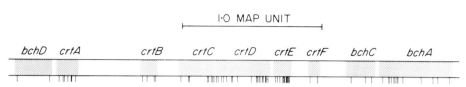

Fig. 9.1 Genetic map of the photopigment region of *Rhodopseudomonas capsulata*. Each shaded area denotes a cluster of mutations that bring about one particular phenotype, and therefore each is probably a gene. See the text for descriptions of each phenotype. Each mark below the line is the map position of a mutation. The ends of many of the genes are not established, especially when only a few mutations have been mapped. Cis-trans tests have confirmed the border between *crtC* and *crtD* and established the *crtD* region as a single cistron.

Map positions of mutations are based solely upon GTA-mediated genetic crosses. One map unit is the amount of DNA carried in a single GTA particle, and therefore mutations separated by one map unit or more are never co-transferred by a single GTA. The map function $d = 1 - \phi^{1/2}$ describes the relationship between distance, d, in map units, and cotransfer frequency, ϕ, the fraction of all transfer events for a given point that also transfer a second point (Yen & Marrs, 1976; Scolnik *et al.*, 1980; and Marrs, unpublished data).

all involved in carotenoid biosynthesis. Those mapped to date are: *crtA*, needed for oxidation of spheroidene to spheroidenone; *crtB*, for an early step of the synthesis of carotenoids, since mutations mapping here block the accumulation of all 40-carbon carotenoids; *crtC*, hydration of neurosporene to hydroxyneurosporene; *crtD*, dehydrogenation of hydroxyneurosporene or methoxyneurosporene to either demethylspheroidene or spheroidene; *crtE*, needed for an early step in carotenoid biosynthesis, since mutations mapping here block the accumulation of all 40-carbon carotenoids; *crtF*, methylation of either hydroxyneurosporene to methoxyneurosporene, or demethyl-spheroidene to spheroidene (Scolnik *et al.*, 1980). The *bch* genes are involved in bacteriochlorophyll biosynthesis. Those mapped to date are: *bchA*, mutants accumulate 2-devinyl-2-α-hydroxyethyl-chlorophyllide *a*, so they may be lacking 'chlorin reductase'; *bchC*, mutants accumulate 2-desacetyl-2-α-hydroxyethyl-bacteriochlorophyllide, and thus they may lack the enzyme responsible for the oxidation of the 2-α-hydroxyethyl group to the 2-acetyl; *bchD*, incapable of phototrophic growth, mutants accumulate greatly reduced amounts of mature bacteriochlorophyll and about one quarter of the normal amount of carotenoids, thus resembling the O_2-repressed state of photopigment production; they may be regulatory mutants. Alternatively the two *bchD* mutants that have been isolated might each have blocks in magnesium chelatase activity that permit a residual low level of activity ('leaky' mutations). Magnesium chelatase-deficient mutants of *Rp. sphaeroides* do not accumulate a detectable intermediate (Lascelles & Hatch, 1969).

The *bchG* locus is required for the conversion of bacteriochlorophyllide to bacteriochlorophyll by the addition of phytol. Mutations in this gene were mapped by *ratio test* crosses. This type of genetic analysis is used to determine the distance between two markers that confer the same phenotype. In this instance distances between *bchD* and *bchG* lesions, both of which block phototrophic growth, were determined. Cells of a strain with a *bchG* mutation were treated with GTA from a strain with a *bchD* mutation, and the number of phototrophically competent recombinants are compared to the number of recombinants obtained for some reference marker. If *bchD* and *bchG* are linked, there will be a reduction in the relative frequency of phototrophically competent recombinants, compared to those obtained in a cross using GTA from a wild type donor. The extent of reduction can be related to the distance between the markers (Yen & Marrs, 1976).

Another genetic locus that has been mapped by ratio test crosses is *rxcA*. Mutations in *rxcA* result in the loss of the three polypeptides that normally form the photosynthetic reaction centres and loss of the light-harvesting I complex as well.

The GTA-based map of the photopigment region is a fine-structure map in that it resolves the relative positions of many mutations within each gene. It currently extends a little more than three map units, which means that it covers about 0.3% of the genome of *Rp. capsulata*. This represents a fairly tight clustering of the genes involved, and suggests that these genes might share common regulatory elements. Many other mutations affecting photosynthesis are known for *Rp. capsulata*, but they did not show linkage to the genes of the mapped cluster or to each other with GTA as the vector. These include *bchE*, *bchH*, *bchF* and *rxcB*. The phenotypes are as follows: *bchE* mutants accumulate P_{590}, which is probably Mg protoporphyrin monomethyl ester; *bchH* mutants accumulate neither bacteriochlorophyll nor precursors that absorb strongly in the visible, and they may lack an enzyme that functions early in the bacteriochlorophyll branch of tetrapyrrole synthesis, such as magnesium chelatase; *bchF* mutants accumulate P_{730}, which may be 2-desacetyl-2-vinyl bacteriochlorophyllide *a*; *rxcB* mutants have the same phenotype as *rxcA*, i.e., they lack reaction centres and light-harvesting I, but they do not map to *rxcA*. The needs for genetic systems capable of mapping these genes by mobilizing larger stretches of DNA and creating partial diploid arrangements led to the search for conjugation. These genes were subsequently mapped using the marker rescue technique described in Section 9.5.2.

9.4.2 Conjugation

The first report of a conjugative genetic system for a phototrophic bacterium was by Sistrom (1977). He transferred an R-factor variant with enhanced sex-factor activity, R68.45, from *Pseudomonas aeruginosa* into *Rp. sphaeroides* by selecting for neomycin-resistance, a marker carried on the plasmid. *Rp. sphaeroides* strains carrying R68.45 can serve as conjugal donors to other strains of this species in a type of mating on solid substrate. All the chromosomal markers tested could be transferred: *lys, met, str, rif, ilv*, and *adn*. The frequencies of these transfer events ranged from about 10^{-4} to 10^{-7} recombinants per donor cell, depending upon the marker transferred. Genetic linkage was demonstrated among several of the markers tested: *str* was linked to *lys, met*, and *ilv*; *rif* was linked to *lys, adn, met*, and *ilv*; *ilv* was linked to *pro*, but not to *met, his*, or *lys* markers; *met* was not linked to *his* or *pro*, nor was *lys*; and the *met-lys* linkage was not tested. Markers related to photosynthesis were not mentioned.

The basis for the enhanced chromosome mobilization ability of R68.45 over R68 seems to be a duplication of a 1.5 Md section of the R-factor DNA (Riess *et al.*, 1980; Leemans *et al.*, 1980). This does not enable R68.45 to mobilize chromosomes from all the hosts in its range, since Wall (personal communication) and Yu *et al.* (1981) found no enhancement for *Rp. capsulata*, and Tucker and Pemberton (1979a) found only low levels of chromosomal mobilization by R68.45 in their strain of *Rp. sphaeroides*. They used strain RS630 and its derivatives, whereas Sistrom used strain WS8, so it would seem that mobilization by R68.45 depends upon some feature that has a more narrow distribution than the host range of the R-factor. We have observed that the 1.5 md duplication that causes chromosome mobilization ability in plasmid pBLM2 is unstable, so negative results with the analogous R68.45 might actually be a result of the reversion of R68.45 back to R68, which has very low chromosome mobilizing ability.

Pemberton and Bowen (1981) have described another chromosome mobilization system for *Rp. sphaeroides*, based on RPL::Tn*501*, and they used that system to construct the first chromosomal-scale map for a phototrophic bacterium. Tn*501* is a mercury-resistance transposon, but it does not alter any of the phenotypic properties of RP1, except chromosome mobilization. Pemberton and Bowen (1981) suggest that perhaps the high frequency with which Tn*501* transposes from one replicon to another accounts for its ability to promote mobilization, whereas Tn*1* and Tn*5* have no parallel effect. Presumably the creation of a region of homology between the chromosome and the plasmid plays a role in the mobilization. While RP1 produces about 10^{-8} recombinants per donor, RP1::Tn*501* produces between 10^{-3} and 10^{-7} depending upon the marker selected. The mating occurs on solid medium. RP1::Tn*501* appears to mobilize the chromosome from two origins, or in two directions from one origin, since two distinct linkage groups are found (Fig. 9.2). Of particular interest is one linkage group that includes a cluster of mutations affecting photopigment synthesis. These genes are clearly linked to typical chromosomal markers, as in *Rp. capsulata*, suggesting that in *Rp. sphaeroides* too the genes for photosynthesis are either chromosomal or episomal. Unfortunately, Pemberton and Bowen (1981) have used the same designations for the mutant alleles they isolated as for those isolated by Sistrom (1977), which may well lead to future confusion. In any event, the

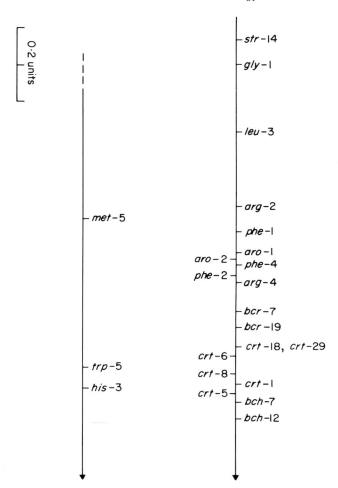

Fig. 9.2 Genetic map of two portions of the *Rhodopseudomonas sphaeroides* genome. The map is based on linkage data obtained from conjugations mediated by RP4::Tn*501*. The map was generated using the empirical formula of Kondorosi *et al.* (1977), $d = 1 - c^{1/3}$, where d is distance in arbitrary units, and c is the linkage frequency between two points. Note that this map unit is much larger than the GTA-based map unit of Fig. 9.1 (From Pemberton & Bowen, 1981.)

linkages observed by Sistrom cannot be compared to the map generated by Pemberton and Bowen because it is not known which alleles are comparable.

The gene cluster specifying photopigment synthesis in *Rp. sphaeroides* is reminiscent of that from *Rp. capsulata*. The *bcr* phenotype is similar to that of *bchD*, both causing lower amounts of pigment synthesis, although *bchD* mutants are more bacteriochlorophyll-deficient than *bcr* mutants. The adjacent genes on the *Rp. sphaeroides* chromosome specify carotenoid biosynthesis as they do in *Rp. capsulata*. The order of genes within the carotenoid region is not precisely analogous; however, the mapping method (Kondorosi *et al.*, 1977) used by Pemberton and Bowen did not order these markers unambiguously. The ability to map and manipulate the genes of *Rp. sphaeroides* should facilitate studies with that organism. However, the abilities to do fine structure mapping and complementation testing are still lacking in this species.

The mobilization of the chromosome of *Rp. capsulata* has been accomplished using a derivative of RP1 with enhanced chromosome mobilization ability (Marrs, 1981). The derivative plasmid, pBLM2, was isolated by screening recombinants that resulted from the low level (10^{-7} per donor) mobilization mediated by pLM2. The latter is a derivative of RP1, with the same low level chromosome mobilization ability as RP1. pBLM2 promotes the formation of up to 6×10^{-4} recombinants per donor and appears to carry the same 1.5 Md DNA insertion as R68.45 (B. Holloway, personal communication), which raises questions about why R68.45 itself did not show chromosome mobilization activity in *Rp. capsulata*. All of the markers tested with pBLM2 were mobilized, including genes for photopigment biosynthesis, cytochrome synthesis, and rifampicin and streptomycin resistance. Linkage among the photopigment markers and between a reaction centre lesion and a carotenoid lesion was very tight (93–99% co-transfer) whereas weaker linkages were observed between photopigment markers and the *trpA* gene (9.4–16% co-transfer) and between a carotenoid lesion and a rifampicin-resistance marker (14% co-transfer).

R-prime derivatives are found among the progeny of pBLM2-mediated crosses at a frequency of about 10^{-6} per donor. They were recognizable in this system because the map of the photopigment region made it possible to construct crosses in which rare progeny could be detected on the basis of colony colour. Cells that were presumed to contain R-primes were found to be genetically unstable, segregating parental and recombinant phenotypes at high frequencies. Presumptive R-primes were mobilized into *E. coli* by plate-mating and selection for kanamycin-resistance. Some of the kanamycin-resistant *E. coli* could be shown to carry R-primes with the genes for photosynthesis from *Rp. capsulata* inserted in the R-factor. The *E. coli* serve as repositories for the R-prime plasmids, since the latter are stable in those cells, and can easily be mobilized into various other strains from them. No expression of the genes for photopigment production has been observed in *E. coli* or *Pseudomonas fluorescens* (Marrs, 1981). The reason for the lack of expression in these other species is unknown, and the question is currently being pursued in Stanley Cohen's laboratory at Stanford University.

When the R-primes are mobilized from either *E. coli* or *P. fluorescens* back to *Rp. capsulata*, a merodiploid condition is established, since two separate copies of the genes for photosynthesis are then present per cell. The partial diploid state is not stable; however, small colonies can form before the instability manifests itself. This process can be visualized by requiring complementation between plasmid and chromosomal genes for pigment synthesis but not for growth. As the colonies grow larger, unpigmented sectors appear near their peripheries. Most often the plasmid-borne genes are eventually lost, but occasionally they replace the chromosomal genes by recombination (Marrs, 1981). Each of the seven independently-isolated R-primes thus far tested carries all of the genes for photosynthetic membrane differentiation for which mutants are known. This includes eight genes affecting bacteriochlorophyll synthesis, four affecting carotenoid synthesis, and two affecting reaction centre synthesis. Not complemented or recombinationally repaired are lesions blocking tryptophan biosynthesis, cytochrome c biosynthesis or cytochrome c-b electron transport. This pattern supports earlier observations about clustering of the genes for photosynthetic membrane differentiation. When complementation is required for growth, colony formation is slow, and tiny colonies are formed. Occasional recombinants that no longer require complementation for photosynthetic growth form large colonies, because the appropriate gene has been integrated into the chromosome. These large colonies are pigmented like either the host, the R-prime donor, or a recombination between the two. Analysis of the frequencies of these events may provide a basis for mapping this region.

The ability to construct merodiploids makes it possible to perform cis-trans complementation tests for defining the boundaries of genes and their dominance relationships. This has been done for the *crtD* region (Serrano & Marrs, unpublished). The genetic map of this region shows two clusters of mutations each conferring the same phenotype (Scolnik *et al.*, 1980). To determine whether each of these clusters represented a separate gene or whether there was one large cistron containing both clusters, three R-primes were constructed which each had one of the following mutations: *crtC76*, *crtD209*, and *crtG83*. The *crtC* mutation served as a positive control, since this gene was known to be a separate cistron. Mutations mapping in *crtC* cause the accumulation of neurosporene, whereas those mapping in the *crtD* and *crtG* clusters both cause the accumulation of a mixture of neurosporene, methoxyneurosporene and hydroxyneurosporene. These three R-primes were each mobilized into a recipient with the *crtD88* mutation by selecting for kanamycin resistant recipients. The *crtD209*–*crtD88* diploid colonies accumulated neurosporene-like carotenoids (green colonies) with rare wild type (red) sectors. The *crtC76*–*crtD88* diploid colonies were mostly red with a minority of green pigment accumulating cells. The *crtG83*–*crtD88* diploids were almost entirely green, indicating that both mutations are in the same cistron, and the 'crtG' region is really just part of the *crtD* gene. This technique can now be used to explore the limits of other genes in this region.

The R-primes, which are transmissible to non-phototrophic species of bacteria, represent recombinant DNA clones which were created *in vivo*. Their

9.4.3 Transformation

Genetic transformation is defined as any process in which naked DNA from a donor is incorporated to effect a genetic change in a recipient. Transformation can be used to analyse the arrangement of genes on chromosomes, or to introduce cloning vectors with recombinant DNA inserts into hosts.

Jasper *et al.* (1978) reported that *Rp. capsulata* could be transformed by colE1::Tn5 DNA, but only at a very low frequency ($\sim 10^{-9}$). The transformation was verified by extraction of plasmid DNA from one kanamycin-resistant transformant, amplification in *E. coli*, and restriction endonuclease analysis, showing that the plasmid colE1::Tn5 was indeed present in the rare transformants. Subsequent attempts to increase the frequency of transformation have been unsuccessful.

Tucker and Pemberton (1980) have described a more useful transformation system for *Rp. sphaeroides*. DNA from bacteriophage Rϕ6P, which carries a gene for a β-lactamase and thus confers penicillin resistance, was introduced into *Rp. sphaeroides* strain RS6143 with the use of a helper phage, Rϕ9. Transformants became lysogenic for Rϕ6P. Rϕ9 is another phage isolated from wild type *Rp. sphaeroides* strain RS901. It appears to be related to Rϕ6P, but the two phages can be distinguished by plaque morphology and immunity. They have the same morphology in the EM, and they both have the unusual covalently closed circular DNA genomes. The highest rates of transformation were observed when RS6143 lysogenic for Rϕ9 was treated with a multiplicity of infection of about 1–10 Rϕ9 per cell at the time of addition of the Rϕ6P DNA. UV irradiation of the Rϕ9 had no effect on its ability to serve as a helper, so the helper effect seems to result from an interaction between the Rϕ9 virions and the cell surface, perhaps aiding in uptake of the circular phage DNA. This system may prove useful for reintroducing cloned fragments of DNA to *Rp. sphaeroides*, especially if Rϕ6P DNA could be used as the cloning vehicle. Helper-phage-mediated transformation might also be a useful means of genetic manipulation for other species of phototrophic bacteria.

The genetic transfer mechanisms described above have been applied to genetic mapping and construction of mutant strains. The helper-phage-dependent transformation system for *Rp. sphaeroides* shows promise for transferring recombinant DNA plasmids, but it has not yet been applied toward that end. A system for mobilization of cloned DNA fragments into *Rp. capsulata* has been developed. It is based on mobilization of colE1-related plasmids by a group Pl R-factor, and it is described in Section 9.5.

In summary, a great deal of progress has been made in the last decade toward developing facile genetic tools for *Rp. capsulata* and *Rp. sphaeroides*.

Genetics for other species of phototrophic bacteria is virtually non-existent. These two species thus offer a distinct experimental advantage for the investigation of most questions concerning bacterial phototrophy.

9.5 GENETIC ENGINEERING AND PROSPECTS FOR FUTURE RESEARCH

9.5.1 Significance

The ability to isolate specific genes and manipulate them *in vitro* has had a revolutionary impact on modern biology. Many long-standing areas of research with the phototrophic bacteria are amenable to approaches based upon DNA cloning. Cloned genes would provide the probes necessary to analyse the patterns of transcription associated with the regulation of photosynthetic membrane synthesis, and thus directly determine the importance and nature of transcriptional control in differentiation of that membrane. The same probes would enable a direct test of the plasmid versus chromosomal location of the genes for photosynthesis or nitrogen fixation. Cloned genes for the pigment-binding proteins could be sequenced. This would allow direct prediction of the amino acid sequences of these hydrophobic proteins which have proven difficult to sequence by standard techniques. Clones of genes for reaction centre and light-harvesting polypeptides would permit mRNAs for those genes to be purified and tested directly for their putative poly A tails (Majumdar & Vipparti, 1980). Clones of genes for photosynthesis would permit comparisons of relatedness of each gene among phototrophic species. The control of transcription of the genes for photosynthesis could be examined *in vitro* using cloned genes as probes for mRNA synthesis, and putative regulatory substances could then be tested directly for their effects on transcription.

Another dimension to the prospects afforded by cloning is the ability to make increased quantities of gene products of interest by *in vitro* alteration of the regulatory elements of the genes. Enzymes or structural proteins could be produced in quantity for further study. They could be produced in the host of origin or in heterologous hosts. It is not unreasonable to think of manipulating the genes for photosynthesis so that they would be functional in a variety of organisms, where they might be more easily studied, or perhaps function in some useful way.

9.5.2 Current cloning capabilities

Although many strategies for cloning the genes for photosynthesis are possible, the only successful system reported to date is based upon the

R-primes isolated from *Rp. capsulata* (Marrs et al., 1980; Clark et al., 1981; Marrs, 1981). The R-prime plasmid pRPS404 is thought to carry all the genes for the photosynthetic membrane differentiation on 39 megadaltons of *Rp. capsulata* DNA integrated with the 36 megadalton pBLM2 R-factor. pRPS404 gives rise to 18EcoRI restriction endonuclease fragments resolvable by agarose gel electrophoresis. Four of these fragments consist mainly of R-factor DNA, and the remaining 14 carry the genes for photosynthesis. Similarly, BamHI digestion gives two R-factor bands and 13 chromosomal bands. Fragments from these sources were ligated into cloning vehicles pDPT44 or pDPT42. These vehicles were derived from plasmid pBR322 *in vitro* by insertion of a piece of DNA bearing a kanamycin-resistance marker, because neither the ampicillin nor the tetracycline-resistance marker of pBR322 is useful in *Rp. capsulata*. pDPT42 has a site suitable for cloning BamHI fragments, and pDPT44 was used for cloning EcoRI fragments. Since small numbers of fragments were involved, collections of *E. coli* clones carrying most of the *Rp. capsulata* restriction fragments were easily established by screening for inserts of the appropriate sizes.

A restriction map showing the relative orientation of the fragments on the *Rp. capsulata* chromosome was generated (Fig. 9.3). This was accomplished by isolating one particular fragment from the BamHI digests, labelling it with ^{32}P by nick translation, and then using the labelled DNA in the Southern transfer and hybridization procedure (Southern, 1975) to determine which EcoRI digestion products overlapped with the particular BamHI fragment. The appropriate EcoRI fragments were then nick translated and used to probe a BamHI digest to establish which BamHI fragments were adjacent to the first. This 'walking' procedure was continued until a large section of DNA, about 20 megadaltons, had been mapped.

In order to identify which genetic determinants were carried on each restriction fragment, it was planned to return each fragment to each of a set of *Rp. capsulata* mutants, and the resulting merodiploid clones would be examined for complementation and recombination. This procedure is called marker rescue. A method for the return of the cloned fragments was developed specifically for this purpose. Plasmids of the P-1 incompatibility group can mobilize some plasmids of other groups in intergeneric matings. The ability to be mobilized requires certain gene products which are not produced by pBR322, but which can be supplied *in trans* by plasmid colE1. P-1 group plasmid R751 was fused to the colE1 derivative RSF2233 to create a plasmid, pDPT55, capable of mobilizing pDPT42 from *E. coli* to *Rp. capsulata*. The fusion between RSF2233 and R751 was necessary because pDPT42 and RSF2233 belong to the same incompatibility group, and RSF2233 would be lost from any cell carrying both plasmids unless RSF2233 were part of another replicon. R751 codes for trimethoprim-resistance, so that *E. coli* strains that acquire pDPT55 can easily be selected. *E. coli* strains bearing both pDPT55 and pDPT42 can conjugate with *Rp. capsulata*, and

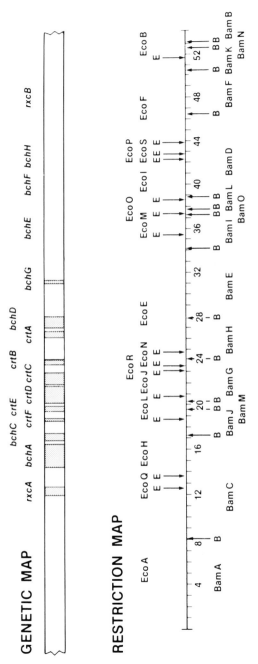

Fig. 9.3 Alignment of the genetic and restriction maps of the region of the *Rp. capsulata* chromosome coding for the photosynthetic apparatus. The *bch* genes affect bacteriochlorophyll synthesis, *crt* genes carotenoid synthesis, and *rxc* reaction centre synthesis. The shaded areas indicate the genetically determined map positions of groups of mutations conferring the same phenotype. The positions of genes for which no shaded area is indicated were determined by marker rescue, but they have not yet been mapped by genetic techniques capable of precise positioning. The 54 kb of *Rp. capsulata* DNA carried in pRPS404 is indicated on the restriction map. Arrows labelled E or B indicate the sites of digestion by EcoRI and BamHI, respectively. The fragments produced by digestion with either enzyme are named alphabetically by size. The junction fragments, EcoA and EcoB and BamA and BamB, carry regions of the vector pBLM2 that are not indicated on this map. EcoC, EcoG and EcoK are composed entirely of DNA from the vector, pBLM2.

pDPT42 can be transferred. The frequency of this mobilization is low (5×10^{-8} per donor) and the progeny are unstable as judged by kanamycin-resistant *Rp. capsulata* exconjugants. This instability proved useful in the isolation of a strain of *Rp. capsulata* that could serve as a more efficient recipient in intergeneric crosses. When spontaneous kanamycin-sensitive segregants from the above cross were retested for recipient ability in the same type of cross (*E. coli* (pDPT55 + pDPT42) × *Rp. capsulata*), one segregant, BY16510, was found that yields 30–1000 times as many kanamycin-resistant exconjugants per donor compared to the parental *Rp. capsulata* strains. BY16510 behaves like a restriction-minus mutant in that while it gives enhanced recipient activity in intergeneric crosses, it is unchanged with regard to crosses with *Rp. capsulata* donors.

Derivatives of BY16510 that carried a variety of mutations affecting the photosynthetic apparatus were isolated and crossed with the fragments of the R-prime pRPS404 cloned in pDPT42. In this way it was learned that the BamG fragment complements the *crtB4* lesion, the BamJ fragment recombines with the *bchC1007* lesion but not *bchC91*, and the BamH fragment recombines with the *bchD1008* lesion but not with *bchD61*. These data, together with the order of fragments given by the restriction mapping and the sizes of the fragments, give a unique colinear alignment between the genetic map and the restriction map. The scaling factor of 1 map unit per 3.0 megadaltons is required to create the alignment. This is aesthetically pleasing, because essentially the same factor was predicted on theoretical grounds based on modelling the map function and the mass of DNA carried in a GTA particle (see Section 9.4.1). Additional marker rescue studies have revealed that the EcoH fragment recombines with *bchA* lesions, the BamE fragment recombines with *bchG* lesions, the BamI fragment carries the *bchE* gene, and the BamD fragment recombines with *bchH* and *bchF* lesions. The *rxcA* and *rxcB* genes are carried by the BamC and BamF fragments, respectively.

Complementation occurred when plasmid pRPSB105, bearing the BamG fragment, was introduced into strains with the *crtB4* lesion, which blocks carotenoid synthesis at an early stage. When the BamG fragment was inserted in the reverse orientation into pDPT42, no complementation was observed, but recombination occurred. This indicates that a promoter located on the vector was responsible for the transcription that permitted expression of the *crtB* gene from pRPSB105. Since the BamG fragment was inserted downstream from the tetracycline promoter of pDPT42, it is presumed that that promoter is responsible. This gives the normal direction of transcription of this region of the photopigment genes as occurring from left to right as drawn in Fig. 9.3.

Complementation also occurs when the cloned BamI fragment is introduced into *Rp. capsulata* cells with *bchE* lesions. In this case, however, the orientation of the insert in the vector does not affect complementation, which occurs in both orientations. This implies that the BamI fragment carries an

Rp. capsulata promoter that is involved in expression of the photosynthetic apparatus.

9.5.3 Prospectus

The availability of cloned fragments with identified genetic content, coding for much of the photosynthetic apparatus, makes the studies proposed in Section 9.5.1 immediately possible, and a time of rapid expansion of our knowledge of phototrophic bacteria is at hand. The *Rp. capsulata* fragments will almost certainly hybridize to those of closely related species, so cloned genes for each of those species should be easily obtainable.

Future genetic engineering involving these organisms could be directed toward utilizing their ability to transduce radiant energy for applied goals. A few examples are discussed below in general terms.

The rhodopseudomonads show much of the metabolic versatility of the pseudomonads which they resemble. The idea of using genetically engineered bacteria to clean up contamination of the environment by organic wastes has been much discussed (Chakrabarty, 1976; Don & Pemberton, 1981). Typical heterotrophs, like the pseudomonads, must obtain both energy and carbon from the organic compounds they degrade, and thus they are theoretically limited in their ability to utilize relatively oxidized compounds. The phototrophic bacteria, in contrast, can afford to expend energy to obtain carbon or nitrogen from such compounds. This ability would seem to make them ideal candidates for suitable engineering. They have already been shown to do an excellent job of reducing biological oxygen demand in a variety of industrial wastes and sewage (Kobayashi & Tchan, 1973). Many of the enzymes involved in catabolism of unusual organic compounds are coded for by plamid-borne genes (Fennewald *et al.*, 1978) and these could be mobilized from the organisms in which they were discovered into strains of phototrophic bacteria for which suitable genetic transfer systems are available.

The energy-harvesting ability of phototrophic bacteria could also be harnessed to produce chemical stocks or fuel supplies. The carotenoids, for example, are polyisoprene compounds which could be converted into a variety of useful organic compounds. It would be a simple matter to delete the gene for oxidation of phytoene, causing the cells to accumulate this relatively reduced carotenoid precursor. Once the regulatory controls of these genes are better understood, it should be possible to turn on the gene for phytoene synthesis to allow for increased synthesis, and more subtle engineering might even allow for extracellular phytoene synthesis.

Another sort of engineering goal would be the improvement of efficiency of certain key enzyme systems. The oxygen sensitivity of the nitrogenase of *Rp. capsulata* might be a target for engineering. This organism has a very active nitrogenase, which might be useful for commercial nitrogen fixation or H_2

evolution, if an inexpensive source of reduced carbon compounds were available (Hillmer & Gest, 1977). By selecting for growth on N_2 in the presence of O_2, mutational changes could be identified that enhance O_2 resistance of the nitrogenase. Strain improvements such as those obtainable by selection on the whole organism, are likely to be difficult to bring together into a single organism, since the underlying mutations may map anywhere and affect any number of aspects of the organism's physiology. If the primary changes do not occur in the nitrogenase itself, then the modifications would obviously not be useful for producing nitrogenase with improved characteristics for use in an abiotic reaction in which purified enzymes or immobilized permeabilized cells catalyse useful reactions.

Some of these difficulties could be circumvented if the target genes were located on a mobilizable plasmid that also carried antibiotic resistance genes. Then the primary selection for oxygen tolerance could be followed by a secondary transfer and selection in a new host. This would select for modifications that occurred in the genes of choice, and would make further *in vitro* alterations easier, such as increasing promoter efficiency or removal of regulatory elements.

Another type of engineering might be the use of phototrophic bacteria to select and modify genes from green plants. A ribulose bisphosphate carboxylase-deficient mutant of a phototrophic bacterium could be isolated by penicillin selection against growth with CO_2 as the sole carbon source. Appropriate mutants might then be used to select for the genes for the carboxylase from green plants, either the chloroplast DNA-coded, catalytic, large subunit or the nuclear-coded, regulatory, small subunit, depending upon the phototrophic bacterium chosen and the nature of the mutant isolated. Once the genes were cloned in a functional state in a bacterial host, the powerful selective techniques available for bacteria could be brought to bear upon the selection of mutants with a lower oxygenase to carboxylase activity ratio, or with an improved K_m for CO_2.

These prospects, speculative as they may seem, probably lie on the conservative end of the spectrum of possible developments in the molecular biology of the phototrophic bacteria.

REFERENCES

BHATNAGAR Y. M. & STACHOW C. S. (1972) Ribosomal proteins of *Rhodopseudomonas palustris*. *J. Bact.*, **109**, 1319–21.

DE BONT J. A. M., SCHLOTEN A. & HANSEN T. A. (1981) DNA-DNA hybridization of *Rhodopseudomonas capsulata*, *Rhodopseudomonas sphaeroides* and *Rhodopseudomonas sulfidophila* strains. *Arch. Microbiol.*, **128**, 271–4.

BREWIN N. J., DE JONG T. M., PHILLIPS D. A. & JOHNSTON A. W. B. (1981) Co-transfer of determinants for hydrogenase activity and nodulation ability in *Rhizobium leguminosarum*. *Nature, Lond.*, **288**, 77–9.

BULL M. J. & LASCELLES J. (1963) The association of protein synthesis with the formation of pigments in some photosynthetic bacteria. *Biochem. J.*, **87**, 15–28.
BURKARDT H. J., RIESS G. & PÜHLER A. (1979) Relationship of group P1 plasmids revealed by heteroduplex experiments: RP1, RP4, R68 and RK2 are identical. *J. gen. Microbiol.*, **114**, 341–8. Chakrabarty A. M. (1976) Plasmids in *Pseudomonas. Ann. Rev. Genet.*, **10**, 7–30.
CHOW C. T. (1976a) Properties of ribonucleic acids from photosynthetic and heterotrophic *Rhodospirillum rubrum. Can. J. Microbiol.*, **22**, 228–36.
CHOW C. T. (1976b) Cell-free, protein-synthesizing system of photosynthetic and heterotrophic *Rhodospirillum rubrum. Can. J. Microbiol.*, **22**, 304–8.
CHOW C. T. (1976c) Functional and structural differences between photosynthetic and heterotrophic *Rhodospirillum rubrum* ribosomes and S-100 fractions. *Can. J. Microbiol.*, **22**, 1522–39.
CHOW C. T. (1977) Multiple forms of DNA-dependent RNA and polyadenylic acid polymerases from heterotrophically grown *Rhodospirillum rubrum. Can. J. Microbiol.*, **23**, 534–58.
CHOW C. T. (1978) DNA-dependent RNA and polyadenylic acid polymerase from phototrophically grown *Rhodospirillum rubrum. Can. J. Microbiol.*, **24**, 1190–6.
CLARK G., TAYLOR D. P., COHEN S. N. & MARRS B. L. (1981) Restriction endonuclease map of the photopigment region of the *Rhodopseudomonas capsulata* chromosome. *Abst. Ann. Mtg. Amer. Soc. Microbiol.*, p. 122.
COHEN-BAZIRE G. & KUNISAWA R. (1963) The fine structure of *Rhodospirillum rubrum. J. Cell. Biol.*, **16**, 401–19.
COHEN-BAZIRE G., SISTROM W. R. & STANIER R. Y. (1957) Kinetic studies of pigment synthesis by non-sulfur purple bacteria. *J. Cell. comp. Physiol.*, **49**, 25–68.
COST H. & GRAY E. D. (1967) Rapidly labeled RNA synthesis during morphogenesis. *Biochim. Biophys. Acta*, **138**, 601–4.
DÉNARIÉ J., ROSENBERG C., BERGERON B., BOUCHER C., MICHEL M., & BARATE DE BERTALMIO M. (1977) Potential of RP4::Mu plasmids for *in vivo* genetic engineering of Gram-negative bacteria. In *DNA Insertion Elements, Plasmids and Episomes* (Ed. by A. I. Bukhari, J. A. Shapiro & S. L. Adhya), pp. 507–35, Cold Spring Harbor Laboratory.
DICKERSON R. E. (1980) Cytochrome *c* and the evolution of energy metabolism. *Sci. Amer.*, **242**, 137–53.
DICKERSON R. E., TIMKOVICH R. & ALMASSY R. J. (1976) The cytochrome fold and the evolution of bacterial energy metabolism. *J. Mol. Biol.*, **100**, 473–91.
DON R. H. & PEMBERTON J. M. (1981) Properties of six pesticide degradation plasmids isolated from *Alcaligenes paradoxus* and *Alcaligenes eutrophus. J. Bact.*, **145**, 681–6.
FENNEWALD M., BENSON S. & SHAPIRO J. (1978) Plasmid-chromosome interactions in the *Pseudomonas* alkane system. *Microbiology*, 1978, 170–3.
FOX G. E., STACKEBRANDT E., HESPELL R. B., GIBSON J., MANILOFF J., DYER T. A., WOLFE R. S., BALCH W. E., TANNER R. S., MAGRUM L. J., ZABLEN L. B., BLAKEMORE R., GUPTA R., BONEN L., LEWIS B. J., STAHL D. A., LUEHRSEN K. R., CHEN K. N. & WOESE C. R. (1980) The phylogeny of prokaryotes. *Science*, **209**, 457–63.
GIBSON J., STACKEBRANDT E., ZABLEN L. B., GUPTA R. & WOESE C. R. (1979) A phylogenetic analysis of the purple photosynthetic bacteria. *Current Microbiol.*, **3**, 59–64.
GIBSON K. D., NEUBERGER A. & TAIT G. H. (1963) Studies on the biosynthesis of porphyrin and bacteriochlorophyll by *Rhodopseudomonas sphaeroides*. 4. S-adenosylmethionine magnesium protoporphyrin methyl transferase. *Biochem. J.*, **88**, 325–33.
GIBSON K. D. & NIEDERMAN R. A. (1970) Characterization of two circular satellite species of deoxyribonucleic acid in *Rhodopseudomonas sphaeroides. Arch. Biochem. Biophys.*, **141**, 694–704.
GRAY E. D. (1967) Studies on the adaptive formation of photosynthetic structures in *Rhodopseudomonas sphaeroides*. I. Synthesis of macromolecules. *Biochim. Biophys. Acta*, **138**, 550–63.

GRAY E. D. (1973) Requirement for protein synthesis for maturation of ribosomal RNA in *Rhodopseudomonas sphaeroides*. *Biochim. Biophys. Acta*, **331**, 390–6.
GRAY E. D. (1978) Ribosomes and RNA metabolism. In *The Photosynthetic Bacteria* (Ed. by R. K. Clayton & W. R. Sistrom), pp. 885–97. Plenum Press, New York.
HAAS D. & HOLLOWAY B. W. (1976) R-factor variants with enhanced sex-factor activity in *Pseudomonas aeruginosa*. *Mol. gen. Genet.* **144**, 243–52.
HIGUCHI M., GOTO K., FUJIMOTO M., NAMIKI O. & KIKUCHI G. (1965) Effect of inhibitors of nucleic acid and protein synthesis on the induced synthesis of bacteriochlorophyll and δ-aminolevulinic acid synthetase by *Rhodopseudomonas sphaeroides*. *Biochim. Biophys. Acta*, **95**, 94–110.
HILLMER P. & GEST H. (1977) H_2 metabolism in the photosynthetic bacterium *Rhodopseudomonas capsulata*: H_2 production by growing cultures. *J. Bact.*, **129**, 724–31.
HU N. T. & MARRS B. L. (1979) Characterization of the plasmid DNAs of *Rhodopseudomonas capsulata*. *Arch. Microbiol.*, **121**, 61–9.
INGRAM V. M. (1972) *Biosynthesis of Macromolecules*, 2e, pp. 118–21. W. A. Benjamin, Inc., Menlo Park.
JASPER P., HU N. T. & MARRS B. (1978) Transfer of plasmid-borne kanamycin-resistance genes to *Rhodopseudomonas capsulata* by transformation and conjugation. *Abst. Ann. Mtg. Amer. Soc. Microbiol.*, p. 114.
DE JESUS T. G. S. & GRAY E. D. (1971) Isoaccepting transfer RNA species in differing morphogenetic states of *Rhodopseudomonas sphaeroides*. *Biochim. Biophys. Acta*, **254**, 419–28.
JOHN P. & WHATLEY F. R. (1975) *Paracoccus denitrificans* and the evolutionary origin of the mitochondrion. *Nature, Lond.*, **254**, 495–8.
KAPLAN S. (1978) Control and kinetics of photosynthetic membrane development. In *The Photosynthetic Bacteria* (Ed. by R. K. Clayton & W. R. Sistrom), pp. 809–39. Plenum Press, New York.
KOBAYASHI M. & TCHAN Y. T. (1973) Treatment of industrial waste solutions and production of useful by-products using a photosynthetic bacterial method. *Water Res.*, **8**, 1219–24.
KONDOROSI A. E., KISS G. B., FORRAI T., VINCZE E. & BANFALVI Z. (1977) Circular linkage map of *Rhizobium meliloti* chromosome. *Nature, Lond.*, **268**, 525–7.
KUHL S. A. & YOCH D. C. (1981) Loss of photosynthetic growth of *Rhodospirillum rubrum* associated with loss of a plasmid. *Abst. Ann. Mtg. Amer. Soc. Microbiol.*, p. 134.
LASCELLES J. (1959) Adaptation to form bacteriochlorophyll in *Rhodopseudomonas sphaeroides*: changes in activity of enzymes concerned in pyrrole synthesis. *Biochem. J.*, **72**, 508–18.
LASCELLES J. & HATCH T. P. (1969) Bacteriochlorophyll and heme synthesis in *Rhodopseudomonas sphaeroides*: possible role of heme in regulation of the branched biosynthetic pathway. *J. Bact.*, **98**, 712–20.
LEEMANS J., VILLARROEL R., SILVA B., VAN MONTAGU M. & SCHELL J. (1980) Direct repetition of a 1.2 Md DNA sequence is involved in site-specific recombination by the P1 plasmid R68. *Gene*, **10**, 319–28.
LESSIE T. G. (1965a) The atypical ribosomal RNA complement of *Rhodopseudomonas sphaeroides*. *J. Gen. Microbiol.*, **39**, 311–20.
LESSIE T. G. (1965b) RNA metabolism of *Rhodopseudomonas sphaeroides* during preferential photopigment synthesis. *J. Gen. Microbiol.*, **41**, 37–45.
MACKAY R. M., ZABLEN L. B., WOESE C. R. & DOOLITTLE W. F. (1979) Homologies in processing and sequence between the 23S ribosomal ribonucleic acids of *Paracoccus denitrificans* and *Rhodopseudomonas sphaeroides*. *Arch. Microbiol.*, **123**, 165–72.
MADIGAN M. T., COX J. C. & GEST H. (1980) Physiology of dark fermentative growth of *Rhodopseudomonas capsulata*. *J. Bact.*, **142**, 908–15.
MADIGAN M., COX J. C. & GEST H. (1981) Photosynthetic pigments in *Rhodopseudomonas capsulata* cells grown anaerobically in darkness. *Abst. Ann. Mtg. Amer. Soc. Microbiol.*, p. 170.

MADIGAN M. T. & GEST H. (1978) Growth of a photosynthetic bacterium anaerobically in darkness, supported by 'oxidant-dependent' sugar fermentation. *Arch. Microbiol.*, **117**, 119–122.

MAJUMDAR P. K. & VIPPARTI V. A. (1980) Polyadenylated messenger RNAs code for photo reaction center and light-harvesting antenna polypeptides of *Rhodospirillum rubrum*. *FEBS Lett.*, **109**, 31–33.

MANSOUR J. D. & STACHOW S. C. (1975) Structural changes in the ribosomes and ribosomal proteins of *Rhodopseudomonas palustris*. *Biochem. Biophys. Res. Comm.*, **62**, 276–81.

MARRS B. (1974) Genetic recombination in *Rhodopseudomonas capsulata*. *Proc. natn. Acad. Sci. USA.*, **71**, 971–3.

MARRS B. L. (1978a) Genetics and bacteriophage. In *The Photosynthetic Bacteria* (Ed. by R. K. Clayton & W. R. Sistrom), pp. 873–83.

MARRS B. L. (1978b) Mutations and genetic manipulations as probes of bacterial photosynthesis. In *Current Topics in Bioenergetics, Vol. 8* (Ed. by D. R. Sanadi & L. P. Vernon) pp. 261–294.

MARRS B. L. (1981) Mobilization of the genes for photosynthesis from *Rhodopseudomonas capsulata* by a promiscuous plasmid. *J. Bact.*, **146**, 1003–12.

MARRS B., WALL J. & GEST H. (1977) Emergence of the biochemical genetics and molecular biology of photosynthetic bacteria. *Trends Biochem. Sci.*, **2**, 105–8.

MARRS B., COHEN S. & TAYLOR D. (1980) Genetic engineering in photosynthetic bacteria. Abstracts of the 5th International Photosynthesis Congress, September 7–13, 1980, Kassandra-Halkidiki, Greece.

MARRS B. & KAPLAN S. (1970) 23S precursor ribosomal RNA of *Rhodopseudomonas sphaeroides*. *J. Mol. Biol.*, **49**, 297–317.

MILLER L. & KAPLAN S. (1978) Plasmid transfer and expression in *Rhodopseudomonas sphaeroides*. *Arch. Biochem. Biophys.*, **187**, 229–34.

NUTI M. P., LEPIDI A. A., PRAKASH R. K., SCHILPEROORT R. A. & CANNON F. C. (1979) Evidence for nitrogen fixation (*nif*) genes on indigenous *Rhizobium* plasmids. *Nature, Lond.*, **282**, 533–5.

OLSEN R. H. & SHIPLEY P. (1973) Host range and properties of the *Pseudomonas aeruginosa* R factor R1822. *J. Bact.*, **113**, 772–80.

PEMBERTON J. M. & BOWEN A. R. ST. G. (1981) High frequency chromosome transfer in *Rhodopseudomonas sphaeroides* promoted by the broad host range plasmid RP1 carrying the mercury transposon, Tn*501*. *J. Bact.*, **147**, 110–17.

PEMBERTON J. M. & TUCKER W. T. (1977) Naturally occurring viral R plasmid with a circular supercoiled genome in the extracellular state. *Nature, Lond.*, **266**, 50–51.

RAZEL A. J. & GRAY E. D. (1978) Interrelationships of isoacceptor phenylalanine tRNA species of *Rhodopseudomonas sphaeroides*. *J. Bact.*, **133**, 1175–80.

RIESS G., HOLLOWAY B. W. & PÜHLER A. (1980) R68.45, a plasmid with chromosome mobilizing ability (Cma) carries a tandem duplication. *Genet. Res. Camb.*, **36**, 99–109.

ROBINSON A. & SYKES J. (1971) A study of the atypical ribosomal RNA components of *Rhodopseudomonas sphaeroides*. *Biochim. Biophys. Acta*, **238**, 99–115.

SAUNDERS V. A. (1978) Genetics of Rhodospirillaceae. *Microbiol. Rev.* **42**, 357–84.

SAUNDERS V. A., SAUNDERS J. R. & BENNET P. M. (1976) Extrachromosomal deoxyribonucleic acid in wild type and photosynthetically incompetent strains of *Rhodopseudomonas sphaeroides*. *J. Bact.*, **125**, 1180–7.

SCOLNIK P. A., WALKER M. A. & MARRS B. L. (1980) Biosynthesis of carotenoids derived from neurosporene in *Rhodopseudomonas capsulata*. *J. Biol. Chem.*, **255**, 2427–32.

SHEPHERD W. D. & KAPLAN S. (1975) Variations in the amounts of three isoaccepting phenylalanyl tRNA species as a function of growth conditions in *Rhodopseudomonas spheroides*. *American Society of Microbiology Abstracts*, p. 102.

SISTROM W. R. (1962) The kinetics of synthesis of photopigments in *Rhodopseudomonas sphaeroides*. *J. Gen. Microbiol.*, **28**, 607–16.

SISTROM W. R. (1963) Observations on the relationship between formation of photopigments and the synthesis of protein in *Rhodopseudomonas sphaeroides*. *J. Gen. Microbiol.*, **28**, 599–605.
SISTROM W. R. (1977) Transfer of chromosomal genes mediated by plasmid R68.45 in *Rhodopseudomonas sphaeroides*. *J. Bact.*, **131**, 526–32.
SOUTHERN E. M. (1975) Detection of specific sequence among DNA fragments separated by gel electrophoresis. *J. Mol. Biol.*, **98**, 503–17.
SUYAMA Y. & GIBSON J. (1966) Satellite DNA in photosynthetic bacteria. *Biochem. Biophys. Res. Commun.*, **24**, 549–53.
TUCKER W. T. & PEMBERTON J. M. (1978) Viral R-plasmid Rϕ6P: properties of the penicillinase plasmid prophage and the supercoiled, circular encapsidated genome. *J. Bact.*, **135**, 207–14.
TUCKER W. T. & PEMBERTON J. M. (1979a) Conjugation and chromosomal transfer in *Rhodopseudomonas sphaeroides* mediated by W and P group plasmids. *FEMS Microbiol. Lett.*, **5**, 173–6.
TUCKER W. T. & PEMBERTON J. M. (1979b) The introduction of RP4::Mu *cts* 62 into *Rhodopseudomonas sphaeroides*. *FEMS Microbiol. Lett.*, **5**, 215–17.
TUCKER W. T. & PEMBERTON J. M. (1980) Transformation of *Rhodopseudomonas sphaeroides* with deoxyribonucleic acid isolated from bacteriophage Rϕ6P. *J. Bact.*, **143**, 43–9.
UFFEN R. L., SYBESMA C. & WOLFE R. S. (1971) Mutants of *Rhodospirillum rubrum* obtained after long-term anaerobic, dark growth. *J. Bact.*, **108**, 1348–56.
WALL J. D., WEAVER P. F. & GEST H. (1975) Gene transfer agents, bacteriophages, and bacteriocins of *Rhodopseudomonas capsulata*. *Arch. Microbiol.*, **105**, 217–24.
WITKIN S. S. & GIBSON K. D. (1972a) Changes in ribonucleic acid turnover during aerobic and anaerobic growth in *Rhodopseudomonas sphaeroides*. *J. Bact.*, **110**, 677–83.
WITKIN S. S. & GIBSON K. D. (1972b) Ribonucleic acid from aerobically and anaerobically grown *Rhodopseudomonas sphaeroides*: comparison by hybridization to chromosomal and satellite deoxyribonucleic acid. *J. Bact.*, **110**, 684–90.
YAMASHITA J. & KAMEN M. (1968) Observations on the nature of pulse labeled RNAs from photosynthetically or heterotrophically grown *Rhodospirillum rubrum*. *Biochim. Biophys. Acta*, **161**, 162–9.
YEN H. C. & MARRS B. (1976) Map of genes for carotenoid and bacteriochlorophyll biosynthesis in *Rhodopseudomonas capsulata*. *J. Bact.*, **126**, 619–29.
YEN H. C. & MARRS B. L. (1977) Growth of *Rhodopseudomonas capsulata* under anaerobic dark conditions with dimethyl sulfoxide. *Arch. Biochem. Biophys.*, **181**, 411–18.
YEN H. C., HU N. T. & MARRS B. L. (1979) Characterization of the gene transfer agent made by an overproducer mutant of *Rhodopseudomonas capsulata*. *J. Mol. Biol.*, **131**, 157–68.
YU P. L., CULLUM J. & DREWS G. (1981) Conjugational transfer systems of *Rhodopseudomonas capsulata* mediated by R plasmids. *Arch. Microbiol.*, **128**, 390–3.

Chapter 10. Evolutionary Roots of Anoxygenic Photosynthetic Energy Conversion

H. GEST

10.1 INTRODUCTION

Discovery of the existence of bacteria that can use light as the source of energy for strictly anaerobic growth, *not associated with O_2 production*, posed one of the most troublesome problems of comparative biochemistry for more than 60 years. Early attempts to demonstrate O_2 formation by photosynthetic bacteria gave negative results (Englemann, 1883) and the 'final' experiments in 1954 (Johnston & Brown) using isotopic techniques did likewise. The extraordinary persistence of the notion that phototrophic bacteria might be capable of producing O_2 had little to do with the lack of suitable experimental techniques to settle the question unambiguously, but rather to mistaken perceptions of the real meaning of comparative biochemistry. For many decades numerous investigators designed experiments with the hope that the results would (must!) reveal a common framework of water cleavage uniting the actual mechanisms of the two types of photosynthesis [see Gest (1982) for an account of the tortuous history of concepts of the comparative biochemistry of photosynthesis].

The desired results never arrived and in retrospect it can now be seen that the wrong questions were being asked on the basis of fragmentary knowledge. Thus, comparative biochemical reasoning went astray for a considerable time. H. A. Krebs (1979) reminds us that research progress is facilitated by asking the right kinds of questions at the right time, and that '... in the past many investigators, including myself, have sometimes failed, either in the design of their experiments or in the interpretation of their results, to consider sufficiently the evolutionary principles regarding biological function, and therefore have missed a great deal.'

The detailed evolutionary connections between anoxygenic and oxygenic photosyntheses are still obscure, but information from several fields leaves little doubt that oxygenic photosynthetic systems were derived from anoxygenic photosynthetic precursors of the kind observed in contemporary green and purple bacteria.

10.2 CYCLIC PHOTOPHOSPHORYLATION: THE CENTRAL FEATURE OF ANOXYGENIC PHOTOSYNTHETIC ENERGY CONVERSION

The possibility that light energy might be convertible to chemical energy in the form of energy-rich phosphate esters was suggested in 1943 by Ruben, but experimental studies during the next several years failed to generate clear-cut supporting evidence. In the meantime, investigators with strong backgrounds in photochemistry and physics argued strenuously, mainly on theoretical grounds, against the possibility that such mechanisms could play a significant role in photosynthesis, [e.g., see Rabinowitch (1945)]. Moreover, an experimental study (Aronoff & Calvin, 1948) with spinach grana, tobacco leaves, and *Chlorella vulgaris* using a radioactive tracer gave negative results: 'Using radioactive phosphorus, no direct connection between *gross* formation of organic phosphorus compounds and photosynthesis or photochemical reductions has been found to occur.' The investigation of this problem using $^{32}P_i$ as a tracer formed the major part of my PhD thesis research with Martin Kamen, and the results (Gest & Kamen, 1948)—with both green algae and *Rhodospirillum rubrum*—showed that illumination caused a marked increase in P_i uptake by intact cells and also a great stimulation in turnover of ^{32}P between soluble and insoluble cell fractions. We concluded that 'The results obtained in this investigation indicate either that phosphorylation is an integral feature of the photosynthetic process proper or that non-related "dark" phosphorylation reactions are greatly stimulated in some way as a consequence of illumination.' Even though the experiments with *Rs. rubrum* were conducted using anaerobic conditions, eliminating the possibility of oxidative phosphorylation, negative feedback inhibition from the Photochemical Establishment apparently led us to add the vague qualifier relating to possible stimulation of a 'dark' process.

The next major advance was made by Frenkel (1954) who discovered that pigmented particles obtained by disruption of *Rs. rubrum* cells can rapidly phosphorylate ADP when illuminated anaerobically in the absence of appreciable quantities of electron donors or acceptors; at about the same time, a similar observation was made by Arnon *et al.* (1954) with green plant chloroplast preparations. Thus, an important common denominator of photosynthetic processes was clearly revealed, setting the stage for the concept of *cyclic* electron flow. In this process, absorption of light by bacteriochlorophyll [Bchl; or some other type of chlorophyll] generates electrons and Bchl$^+$. The electrons are transferred anaerobically via appropriate carriers to Bchl$^+$, thus completing the cycle (illustrated in Fig. 10.1). In essence, light creates an electron current that drives the phosphorylation of ADP without the necessity for external sources of electrons or terminal oxidants.

From the research of numerous investigators, we can conclude that whatever metabolic idiosyncrasies they may show in respect to carbon

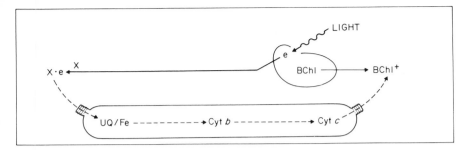

Fig. 10.1 The cyclic pathway of light-dependent electron flow as observed in typical phototrophic bacteria. Bacteriochlorophyll (BChl) and the various electron carriers (X, the primary electron acceptor; UQ/Fe, ubiquinone-iron complexes; Cyt b, cytochrome b; Cyt c, cytochrome c) are embedded in cellular membranes.

metabolism, electron sources for reductive biosyntheses etc., all phototrophic bacteria have at least one major property in common—the ability to catalyse a light-dependent cyclic electron flow that drives synthesis of ATP from $ADP + P_i$. As in all other systems in which electron flow provides the energy for phosphorylation of ADP, 'photophosphorylation' is associated with catalysts embedded in cellular membranes. It is convenient to use the general term 'electrophosphorylation' to describe such processes (Gest, 1980). Accordingly, the cyclic photophosphorylation observed in phototrophic bacteria can be considered to be a type of anaerobic electrophosphorylation in which exogenous reductants or oxidants are unnecessary. Can a plausible scheme be formulated for the origin of an energy conversion mechanism of this kind, in part at least, from 'earlier' kinds of non-photosynthetic energy transduction? The vast knowledge of prokaryotic physiology and metabolism now available provides the possibility for suggesting a reasonable scenario.*

10.3 EARLY BIOENERGETICS; FERMENTATION

As a starting point, I state the familiar assumption that the earliest cell lines obtained energy from (anaerobic) fermentative metabolism. There are a number of cellular indicators that sugars were especially important in this connection. Thus, the special significance of sugars in biochemical evolution has been aptly summarized by Quayle and Ferenci (1978) as follows: 'The occurrence of sugars in the macromolecules (nucleic acids, polysaccharides,

---

* Scenario: 'a summary or outline of the plot of a dramatic work.' The 'invention' of photosynthetic metabolism was unquestionably one of the most dramatic events in biochemical evolution.

membranes, cell walls, etc.) and intermediary metabolites of all known cells attests to the importance of sugars in the earliest development of self-replicating organisms.'

Glucose has been described as the universal cellular fuel. Indeed, the capacity for fermentative utilization of glucose (or related sugars) as an energy source undoubtedly is the most ubiquitous of the biological energy-conversion mechanisms known in extant organisms. This is consistent with the view (Oparin, 1938) that the earliest organisms were fermenters. Various arguments can be made for believing that sugar fermentation was the first energy-conversion process successfully exploited by cells growing in the primitive anaerobic atmosphere. Thus:

1. 'Thermodynamic stability can be invoked, for example, for the selection of glucose as the universal cellular fuel. Glucose, the most stable of the 16 isomeric hexoses with its bulky substituents in equatorial positions, had the greatest chance to accumulate in the primordial soup' (Bloch, 1979).
2. Fermentation of glucose, as exemplified by glycolysis to lactic acid, is mechanistically the simplest known type of energy conversion, and requires only one kind of hydrogen (electron) carrier (viz., NAD).
3. ATP generation in fermentations occurs via 'substrate-level' phosphorylation in completely soluble systems; thus, membranous cellular structures are not required.
4. The yield of ATP from fermentation of sugars is quite low, as would be expected of an early, inefficient energy-conversion mechanism.
5. 'The view that anaerobic fermentations were the earliest energy-supplying processes is consistent with the finding that some of the components of the fermentation reactions, including ATP and DPN [NAD], occur in almost every other metabolic process supplying, or depending on, utilisable energy' (Krebs & Kornberg, 1957).

What was the source of sugars in the 'primeval soup'? Quayle and Ferenci (1978) give an excellent summary of ideas relating to the possible importance of formaldehyde (HCHO) in the prebiotic environment, and of experiments on the abiotic self-condensation of HCHO to sugars (the formose reaction). They suggest that upon exhaustion of the pool of prebiotic sugars on the primitive earth, the shortage of sugar was overcome by net synthesis from HCHO by a pathway similar to the ribulose monophosphate (RuMP) cycle now observed in many methylotrophs (bacteria that can use reduced C_1 compounds as sole sources of carbon). It is noteworthy that carbon transformations in the RuMP cycle are strikingly similar to those of the ribulose bisphosphate (Calvin) cycle of CO_2 fixation, and it has been suggested that the latter evolved from the former (Quayle & Ferenci, 1978; see also Quayle, 1980).

10.4 EVOLUTION OF FERMENTATION MECHANISMS

To most biochemists, fermentation suggests only the classical energy-yielding mechanisms by which glucose is anaerobically converted to lactic acid, or ethanol plus CO_2. This is probably due in large measure to the fact that elucidation of these particular fermentations constituted one of the major thrusts of modern biochemistry. The processes noted, however, are relatively simple in comparison with the fermentation patterns of typical saccharolytic clostridia and various other anaerobes that produce a multitude of extracellular end-products from sugars.

What is the significance of the various complex fermentation patterns observed in contemporary anaerobes? In my view, these represent the results of evolutionary 'experiments' aimed at solving the problem of achieving oxidation-reduction (O/R) balance in the most propitious way. The Embden-Meyerhof-Parnas (EMP) pathway can be considered to be the prototype of fermentation, and in this mechanism 2 $NADH_2$ must be reoxidized to regenerate the NAD required for continued fermentation. In the 'classic' fermentations this occurs simply by reduction of organic fermentation intermediates, notably pyruvate or an immediate derivative such as acetaldehyde or acetyl-Coenzyme A.

The disposal of H from $NADH_2$ can be seen as a crucial feature in the sugar metabolism of growing fermentative anaerobes from the following argument. Cell yields of anaerobes growing on glucose are usually not more than 20 g dry weight per mole of sugar fermented. If the organism is using the EMP pathway and produces lactic acid as the end product, this is tantamount to saying that per mole of glucose, 4 g of H are mobilized during the fermentation (2 $NADH_2$/glucose fermented) and must be disposed of. Since the H content of dry cell material is usually ca. 7%, in our example only $20 \times 0.07 = 1.4$ g H can be used for cell synthesis. This leaves 2.6 g H to be accounted for. Thus, pyruvate (or an immediate derivative) or an external oxidant must be reduced in order to achieve redox balance. The imbalance noted is exacerbated by other factors which Krebs (1972) has drawn attention to in a discussion on the relations between respiration and fermentation:

> 'So at best anaerobic fermentation may be expected to replace the aerobic ATP supply, provided that the rate of fermentation is high enough, but there are other features of respiration which fermentation itself cannot replace: it cannot maintain the internal environment of the cell in respect to the redox state of the pyridine nucleotides and other redox catalysts. For this the great majority of cells require a continuous supply of oxygen. Micro-organisms which can grow anaerobically in simple media have evolved special mechanisms for the maintenance of the redox state. The energy supply by lactic acid or alcoholic fermentation is neutral with respect to changes of the redox

balance, since oxidations and reductions are equal. However, when cells grow on a simple medium with glucose as the sole source of carbon they have to synthesize many cell constituents by reactions which involve an excess of oxidations over reductions. Only a small minority of syntheses involve an excess of reductions while some biosyntheses are balanced with respect to oxidations and reductions.'

If the disposal of H from $NADH_2$ in a fermentation is obligatorily coupled to pyruvate reduction, this important intermediate [or its precursor phosphoenolypyruvate (PEP)] becomes unavailable for biosynthesis. I envisage that one of the early tides of metabolic evolution consisted of 'trials' to improve efficiency by minimizing the use of pyruvate simply as an H (electron) 'dump.' It is reasonable to assume that fermentative metabolism gradually evolved so as to give: (a) increasing yields of ATP per molecule of substrate fermented, and (b) more efficient, flexible use of fermentation intermediates for major biosynthetic purposes. Improvements in (a) and (b) need not have occurred in concert in all developing cell lines, and many of the evolutionary experiments aimed at greater efficiency have probably survived among the many species of contemporary heterotrophic anaerobes that show specialized fermentation mechanisms.

10.4.1 Accessory oxidant-dependent fermentations

The most familiar fermentations can be regarded as being internally balanced from the standpoint of O/R, that is, the terminal electron acceptor for the energy-yielding O/R reaction of the fermentation mechanism is generated from the fermentation substrate. Do fermentations exist that are balanced externally, i.e., by reduction of an oxidant *not* derived from the organic fermentation substrate? F. Egami and his colleagues were very prescient in suggesting, in 1957, that certain kinds of clostridia (as well as other systems) may use inorganic nitrate in this fashion. Thus, they proposed the term 'nitrate fermentation' to describe sugar fermentation in which $NADH_2$ is recycled to NAD by H transfer to a soluble system that reduces nitrate to nitrite. This interesting idea of inorganic fermentations, supported at the time only by rather indirect experimental observations, was interpreted in subsequent papers (see Egami, 1974 and Gest, 1981b) as an important stage in the early evolution of energy-transducing systems. [Note that Cole & Brown (1980) have described experiments indicating that anaerobic reduction of nitrite to ammonia during sugar fermentation by several facultative anaerobes results from *non-phosphorylative* oxidation of $NADH_2$ with nitrite as the electron acceptor.]

It seems inescapable that biological energy conversion mechanisms must have evolved from relatively inefficient anaerobic fermentations to more efficient processes, and that this occurred through a lengthy sequence of gradual and interdependent improvements in both bioenergetic and biosynthetic capacities. This consideration underlies the author's recent theory (Gest, 1980) on the evolution of fermentation mechanisms. In essence, it is suggested that the earliest fermentations were internally balanced in respect to O/R and became altered through the employment of accessory oxidants, thus sparing pyruvate (or PEP) for biosynthetic use. Although nitrate could, in principle, serve the function of an accessory oxidant, the theory is modelled on exploitation of CO_2 for this purpose.*

10.4.2 Use of CO_2 in biosynthesis of accessory oxidants

Carbon dioxide, the most oxidized form of carbon, performs the function of an electron sink in the metabolism of a large number of anaerobes and facultative anaerobes. In many instances, this occurs through the condensation of CO_2 with a C_3 intermediate of fermentation (either pyruvate or PEP) to yield a relatively oxidized C_4 dicarboxylic acid, oxaloacetate (OAA) (Eqn. 10.1); OAA and its derivatives can serve as electron sinks for 4 H atoms.

$CO_2 + {}'C_3{}' \rightarrow HOOCCOCH_2COOH$ (oxaloacetate)   **Eqn. 10.1**

$\qquad\qquad\qquad +2H \downarrow$

$\qquad\qquad HOOCCHOHCH_2COOH$ (malate)

$\qquad\qquad\qquad \downarrow -H_2O$

$\qquad\qquad HOOCCH=CHCOOH$ (fumarate)

$\qquad\qquad\qquad +2H \downarrow$

$\qquad\qquad HOOCCH_2-CH_2COOH$ (succinate)

Some bacteria require substantial quantities of exogenous CO_2 in order to ferment sugars and this has been attributed to the role of CO_2 as a precursor of OAA (and thus, fumarate; Anderson & Ordal, 1961; White et al., 1962). In other instances, exogenously added fumarate, by acting as an H acceptor, permits fermentative growth of bacteria on lactate (Quastel et al., 1925) or glycerol (Singh & Bragg, 1975). Accordingly, fumarate can be considered as another example of an accessory oxidant that facilitates redox balance in fermentative metabolism.

Depending on the organism (or tissue in higher eukaryotes) the C_3 precursor of OAA can be either pyruvate or PEP. From the standpoint of

---

* Fermentations in which redox balance is facilitated by reduction of other kinds of oxidants, and evolution of H_2 are discussed by Gest & Schopf (1983).

oxidation-reduction, the net result is the same in both instances. The special importance of the sequence depicted above is that it represents a mechanism for reoxidation of the $NADH_2$ generated in the EMP pathway through utilization of *only one* of the C_3 fragments produced from hexose cleavage. Operation of the 'C_4 dicarboxylic acid electron sink sequence' has the consequence that the other C_3 fragment is potentially available for biosynthesis. This conception is illustrated in condensed form in Fig. 10.2. Since pyruvate (or PEP) generally plays a central role in biosynthetic mechanisms, it is reasonable to assume that an anaerobic cell with the capacity to exploit CO_2 for the synthesis of C_4 electron acceptors would have a significant selective advantage over cells constrained to use a hexose as an energy source only by the classical glycolytic mechanism (Fig. 10.2, top). Thus, it is envisaged 'that in its earliest form, fumarate reduction evolved as a means of achieving redox balance and at the same time sparing pyruvate for biosynthetic purposes.' (Gest, 1980).

The sequence: hexose $\longrightarrow C_3 \xrightarrow{CO_2} OAA \xrightarrow{+4H}$ succinate which has been

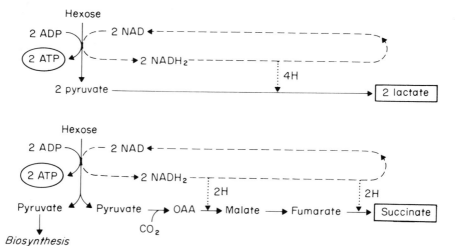

Fig. 10.2 (Top) Condensed scheme for energy-yielding hexose fermentation by the Embden-Meyerhof-Parnas pathway. The fermentation is 'internally balanced' from the standpoint of oxidation-reduction; NAD is regenerated from $NADH_2$ by transfer of H atoms to the intermediate pyruvate, resulting in the accumulation of lactate. (Bottom) A variation of hexose fermentation in which reoxidation of $NADH_2$ is effected by H transfer to the dicarboxylic acids, oxaloacetate (OAA) and fumarate. The latter is the terminal H acceptor, yielding succinate. Carboxylation of pyruvate or a related C_3 compound gives rise to the relatively oxidized C_4 dicarboxylic acids. In many present-day organisms, phosphoenolpyruvate rather than pyruvate is the actual C_3 moiety that condenses with CO_2 to form oxaloacetate; this has no effect on the redox balance which is the central point at issue. It is suggested that this pattern of redox balance spares pyruvate for use in biosynthetic metabolism.   (from Gest, 1980)

designated the 'Szent-Györgyi pathway' (Ottaway & Mowbray, 1977), is of particular interest for investigators of biochemical evolution. In addition to the electron sink function of the $C_3 \longrightarrow$ succinate segment, the reductive sequence from OAA to succinate can be construed as the evolutionary starting point for the citric acid cycle of heterotrophic aerobes, a central feature of their energy metabolism. In such aerobes, the sequence in question runs in the oxidative direction, i.e., succinate $\longrightarrow$ fumarate $\longrightarrow$ malate $\longrightarrow$ OAA. Before O_2 appeared on the Earth in quantity, the same sequence presumably operated in the *reductive* direction as an electron sink and also to provide succinate (succinyl-CoA) needed for biosynthesis. In most aerobes, the product of α-ketoglutarate dehydrogenase activity, succinyl-CoA, is a branch-point in the sense that it is converted to succinate (the next intermediate of the citric acid cycle) and also is a biosynthetic precursor of metalloporphyrins. Anaerobes that produce metalloporphyrins (e.g., cytochrome b) and lack α-ketoglutarate dehydrogenase apparently produce succinyl-CoA needed for biosynthesis by the sequence: OAA $\longrightarrow$ malate $\longrightarrow$ fumarate $\longrightarrow$ succinate $\longrightarrow$ succinyl CoA.

Ubiquity of a metabolic system argues for antiquity, and the widespread distribution of the capacity to use fumarate as an electron sink is striking. Numerous and diverse prokaryotes show this ability (see listings in Thauer *et al.*, 1977 and Kröger, 1977) and, in addition, the Szent-Györgyi pathway is a prominent aspect of metabolism of many invertebrates that can survive long periods in essentially anaerobic habitats, (e.g., see Hochachka & Mustafa, 1972; Hochackha, 1976).

10.4.3 Evolution of the fumarate reduction system

Effective utilization of fumarate as a terminal electron acceptor depends on the activity of a reductase with appropriate kinetic properties. The existence of an essentially unidirectional fumarate reductase was first clearly recognized in studies with the strict anaerobe *Micrococcus lactilyticus* [now known as *Veillonella alcalescens*; Peck *et al.*, (1957)]. Extracts of this bacterium were shown to contain a soluble, flavin-linked fumarate reductase activity that clearly could not be ascribed to conventional succinate dehydrogenase acting in the reverse direction. In subsequent research with a variety of anaerobes and facultative anaerobes it has been found that fumarate reductase (FR) and other components of the fumarate reduction system are almost always membrane-bound (Thauer *et al.*, 1977). The sequence of steps in electron transport to fumarate in different organisms is not always identical, as might be expected, but generally follows the pattern:

electron donor $\longrightarrow$ menaquinone $\longrightarrow$ cytochrome $b \longrightarrow$ (FR) $\begin{array}{l}\nearrow \text{Fumarate} \\ \searrow \text{Succinate}\end{array}$

NADH$_2$, generated in the course of hexose fermentation, serves as an electron donor in a number of instances, and electrophosphorylation has been demonstrated with such systems (Thauer et al., 1977; Kröger, 1978). These considerations formed part of the basis of the scheme advanced by Gest (1980) for evolutionary changes of fermentation mechanisms (Fig. 10.3).

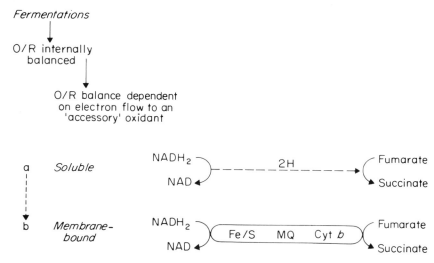

Fig. 10.3 Proposed sequence of evolutionary change of fermentation schemes. Fermentation in which oxidation-reduction (O/R) is internally balanced as in Fig. 10.2, top, is modified to the pattern in Fig. 10.2, bottom, at first as in (a), where reoxidation of NADH$_2$ is effected by a soluble, relatively simple system in which fumarate is the terminal oxidant. Through sequential changes, the accessory oxidant system becomes more complex with Fe/S proteins, menaquinone (MQ), and cytochrome b (Cyt b) participating as catalysts in a membrane-bound system capable of electrophosphorylation (b).   (from Gest, 1980)

It is envisaged that the earliest (simplest) fermentations were internally balanced from an O/R standpoint, as in classical glycolysis (hexose ⟶ 2 lactate). In a later stage, fermentations were modified so as to spare pyruvate (or PEP) for biosynthesis, through acquisition of H transfer pathways to accessory oxidants (another advantageous aspect of sparing use of pyruvate simply as a terminal H acceptor was that pyruvate can be converted to acetyl-CoA, a potential source of additional energy). Early accessory oxidant systems used for reoxidation of NADH$_2$ presumably were relatively simple and soluble, as indicated in Fig. 10.3a, and functioned only to ensure redox balance. The ubiquity of the fumarate-reducing system and the obviously close connections of the C$_4$ intermediates of the Szent-Györgyi pathway with other aspects of metabolism strongly imply that the use of fumarate as a terminal oxidant was an important prototype for evolutionary modification of fermentations. The scheme of Fig. 10.3 suggests that further biochemical evolution involved gradual addition of intermediary electron carriers (Fe/S proteins, menaquinone, etc.) and incorporation of most of the system into the cytoplasmic membrane. The sequence as depicted in Fig. 10.3b essentially represents the membrane-bound system as it now exists in numerous organisms, including strict anaerobes. Thus, an electrophosphorylation 'module' is found in organisms that can obtain the bulk of their ATP by conventional fermentation.

It is not difficult to imagine early prokaryotes in which sugar was fermented by mixtures of ordinary glycolysis and modifications as in the Szent-Györgyi pathway. In any event, the appearance of an electrophosphorylation module in fermenters can be seen as a major step forward, pointing towards the later development of even more efficient systems in which enhanced electron flow in membranes was the driving force of energy conversion.

10.4.4 Accessory oxidant-dependent fermentation in phototrophic bacteria

During the 1930s several investigators noted the endogenous production of gases and acidic products by resting cells of various types of (phototrophically-grown) purple bacteria incubated in darkness under anaerobic conditions. The acidic products were not identified, and it was claimed that their formation could not be appreciably enhanced by addition of organic substrates. These observations, together with the fact that no one had been able to grow such organisms anaerobically in darkness, led to the view that the fermentative metabolism of phototrophic bacteria was of a peculiar 'special' nature, and possibly functioned only for 'maintenance' of the cells during anaerobic dark periods. Later experiments with *Rs. rubrum* showed, however, that resting cells of this organism could ferment pyruvate anaerobically in darkness by a mechanism similar to that operating in heterotrophic propionic acid bacteria (Kohlmiller & Gest, 1951). Since the latter organisms *can* obtain the energy and carbon required for growth from pyruvate fermentation, it was concluded (Kohlmiller & Gest, 1951; Gest, 1951) that under appropriate conditions, photosynthetic bacteria should also be able to grow fermentatively in darkness. Experimental trials, however, continued to give negative results until Uffen and Wolfe (1970) reported that several phototrophic bacteria (including *Rs. rubrum*) could be grown in complex media (containing yeast extract and peptone) under anaerobic conditions in darkness when special procedures were used to establish strict anaerobiosis and a low E_h.

A number of photosynthetic bacteria can utilize sugars as sole carbon sources for anaerobic *photoheterotrophic* growth, but common experience over many years indicated that sugars could not support dark anaerobic multiplication. It was, consequently, surprising when Yen and Marrs (1977) described experiments which demonstrated that *Rhodopseudomonas capsulata* and closely related species can grow anaerobically in darkness in synthetic media with glucose as the sole carbon and energy source provided that dimethyl sulphoxide (DMSO) is supplied. During growth, the DMSO is reduced to dimethyl sulphide, leading Yen and Marrs to suggest that under the conditions described, ATP is produced either in association with an anaerobic respiration in which DMSO functions as a terminal electron acceptor, i.e., electrophosphorylation, or by an unusual fermentation (substrate-level phosphorylation) process that requires an accessory oxidant. More detailed studies using trimethylamine-N-oxide (TMAO) as the oxidant in place of DMSO have established that the energy-yielding process involved is indeed a fermentation dependent on an accessory oxidant (Madigan & Gest, 1978; Madigan *et al.*, 1980; Cox *et al.*, 1980).

What is the function of the accessory oxidant? *Rp. capsulata* apparently contains virtually all of the classical fermentation machinery found in many heterotrophic anaerobes, but differs in that it cannot effectively manage O/R balance using only fermentation intermediates as terminal electron acceptors. For reasons still unknown, during anaerobic dark growth the organism is unable to reoxidize $NADH_2$ generated in the EMP pathway by making highly reduced organic end-products (or H_2) and an exogenous accessory electron acceptor (TMAO or DMSO) must therefore be provided. The essence of this

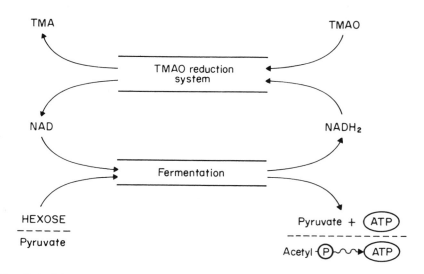

Fig. 10.4 Schematic representation of the dark fermentative metabolism of *Rhodopseudomonas capsulata*. Energy-yielding catabolism of hexose to pyruvate and the conversion of pyruvate to acetyl phosphate (a phosphoryl donor for ATP synthesis) require NAD as a H (electron) acceptor. In order for fermentation to proceed, NAD must be continuously regenerated from $NADH_2$. This is accomplished by transfer of H to the accessory oxidant trimethylamine-N-oxide (TMAO), which is thereby reduced to trimethylamine (TMA). (from Madigan *et al.*, 1980)

interpretation is shown in Fig. 10.4. As indicated, NAD is continually supplied to the fermentation apparatus by transfer of H from $NADH_2$ to the TMAO-reduction system. The latter functions simply as a dumping-ground, or sink, for electrons [Cox et al. (1980) give evidence showing that electron flow from $NADH_2$ to TMAO is catalysed by a soluble non-phosphorylating enzyme system.]

The discovery in a phototrophic bacterium of a cryptic fermentation system evoked only by adding an accessory oxidant raises many questions of interest. Perhaps this kind of mechanism in *Rhodopseudomonas* is an evolutionary relic from a lengthy era in which fermentations were the only energy-conversion devices available to prokaryotic cells, which were undergoing various evolutionary changes in the direction of greater efficiency in the cellular metabolic economy.

10.5 A MODEL FOR THE ORIGIN OF ANAEROBIC PHOTOPHOSPHORYLATION

In a recent review, F. Harold (1978) remarked: 'It is quite striking that the constellation of nonhaem iron-quinone-cytochrome *b*, however it may have arisen, was strongly conserved thereafter. It is a common feature of the redox chains of both photosynthetic and respiratory organisms, and it will be very interesting to see whether it also occurs in anaerobes such as sulphate reducers, which are thought to be an ancient group.' As noted earlier, the constellation Harold refers to is known to be commonly found in anaerobes and amphiaerobes.* The close similarities of the electrophosphorylation module linked with fumarate reduction and the electron carrier chain used in light-dependent energy conversion by extant purple phototrophic bacteria suggested, in part, the hypothesis proposed by the author (Gest, 1980) for the origin of anaerobic photophosphorylation. As indicated in Fig. 10.5, the essence of the scheme is that the first photophosphorylation system was established through fusion of an electrophosphorylation module in a fermentative anaerobe with a membrane-associated photopigment (Mg porphyrin) complex. Accordingly, the main evolutionary roots of anoxygenic photophosphorylation were a module from a relatively complex accessory oxidant-dependent fermentation and a photoreactive Mg porphyrin. It is considered likely that Mg porphyrins were first produced biologically through modification of the pathway established for synthesis of iron porphyrins (Calvin, 1959). Mg porphyrins, of which Bchl is an example, are unusually reactive from a photochemical standpoint. In order for the photoproducts to

---

* An amphiaerobe is defined as an organism or cell that can use O_2 as the terminal electron acceptor for energy conversion or some alternative energy transduction process that is independent of O_2 (Chapman & Gest, 1983).

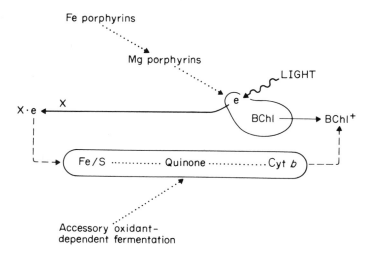

Fig. 10.5 Scheme for origin of anaerobic photophosphorylation. Synthesis of a form of bacteriochlorophyll (BChl; not necessarily identical in structure to the chlorophylls of present-day phototrophic bacteria) occurred through modification of the pathway for biosynthesis of iron porphyrins. The charge separation induced by the absorption of light became metabolically useful only after the photopigment system was stabilized by fusion with an electrophosphorylation-competent segment of an accessory oxidant-dependent fermentation mechanism. X represents the primary electron acceptor of the photosynthetic apparatus. (from Gest, 1980)

be useful metabolically, however, a means must be provided for stabilizing the light-induced separation of charges. Unless there is a mechanism for stabilization, the charged entities created by light absorption recombine very rapidly and the energy potentially available is dissipated (Krasnovskii, 1959; see also Chapter 3).

The widespread occurrence of the fumarate-reducing electrophosphorylation module and the close relationships of fumarate and succinate to other aspects of metabolism suggests that the electrophosphorylation catalysts of this particular system may have been recruited for the first successful fusion. After photophosphorylative energy conversion was established in a cell line, the continued operation of a relatively inefficient accessory oxidant-dependent fermentation can be seen as becoming gratuitous. In this connection it may be noted that fumarate cannot substitute for TMAO as an accessory oxidant for dark fermentative growth of *Rp. capsulata* (Madigan & Gest, 1978).

10.5.1 Enslavement of photochemistry by biochemistry

A number of known genetic recombination or exchange mechanisms could have led to a fusion of the kind suggested in Fig. 10.5, and further speculation in this connection is not particularly useful at present. A more relevant question is: how were the extraordinarily rapid processes of quantum absorption coupled to the comparatively very slow reactions of cellular chemistry? Gaffron (1965) described the overall result of the coupling as follows:

> 'Students of sensitized photochemical reactions *in vitro* know of course that the greatest obstacle to a practical utilization of light energy (in this manner) is the tendency of the reaction products to recombine either directly or via a short detour of at most one or two intermediate steps... The living cell has solved the problem of premature and useless back reactions by making the detour longer and longer; in other words, by converting the back reactions into a cycle to which other metabolic processes can be coupled...'

Stated in another way, a mechanism was acquired that stabilized light-induced separation of charges and facilitated an alternative route of step-wise electron transfer, geared to conservation of energy in forms utilizable for biosynthesis.

Electron transfer catalysts in electrophosphorylation modules necessarily have an appropriate vectorial arrangement in membranes (Mitchell, 1979). For a successful mechanistic fusion of such a module with a photopigment complex, the latter would have to have assumed a compatible topological arrangement. One can imagine many trials and errors before this was achieved. The resulting photosynthetic unit in which a photoactivatable pigment could replace organic compounds as an electron source for energy conversion was a signal development linking biological systems directly to an inexhaustible energy source. Further evolutionary improvement can be seen in addition of a *c* type cytochrome to the sequence of electrophosphorylation catalysts and later, coordination with a second photopigment-protein system which made possible the utilization of water as an electron source for light-dependent electrophosphorylation and generation of net reducing power, i.e., oxygenic photosynthesis.

This model for the origin of photosynthetic metabolism suggests new possibilities in tracing biochemical evolution through analysis of the biochemistry and molecular biology of extant organisms. In particular, it would be of special interest to explore the fine structure of accessory oxidant-dependent fermentations in a variety of organisms, including all the major kinds of phototrophic bacteria. The fusion model presented for the origin of anoxygenic photosynthetic energy conversion contains elements suggested in very general fashion by others. For example, Krasnovskii (1959) described the early evolution of photosynthetic systems as follows:

> 'The system of heterotrophic metabolism in ancient organisms which were not yet capable of light utilization apparently included the steps

of catalytic electron transfer and "anaerobic" oxidoreductive phosphorylation. The development of the pigment system had the result that the active products of the photoprocess were linked to formerly existing biocatalytic systems. The further development of catalytic pigment and systems led to the more rational utilization of light energy.'

The main thrust of the more detailed scheme described here stems from the conviction that the very remarkable intermeshing of bioenergetics and biosynthesis in all types of contemporary cells indicates that these two interdependent aspects of cellular biochemistry must have evolved in a delicately concerted way. Accordingly, the progression from the earliest anaerobic cells with simple fermentative energy conversion and limited biosynthetic ability to cells with high electrophosphorylation capacity and the biosynthetic competence to make all cellular constituents from a single organic compound, e.g., a hexose, must have occurred through a lengthy sequence of gradual and essentially simultaneous improvements in both biosynthetic and bioenergetic mechanisms.

Previous speculations which attempted analysis of *cellular* evolution based almost entirely on consideration of only bioenergetic mechanisms obviously must give distorted glimpses of an incredibly complex tide of biochemical evolution. Similarly, current attempts to construct prokaryotic evolutionary trees based on sequence analyses of particular proteins or nucleic acids have led to implausible suggestions of close evolutionary relationships between organisms that have very different physiological and biochemical characteristics (in this connection see comments by Demoulin, 1979).

10.6 A SYNOPSIS OF THE CONSEQUENCES OF INVENTION OF ANAEROBIC PHOTOPHOSPHORYLATION

The first successful assembly of a mechanism for anaerobic conversion of light energy to chemical energy (cyclic photophosphorylation) utilizable for biosynthesis can be viewed as a major spring-board toward the further evolution of bioenergetic systems of increasingly higher efficiency. The selective advantage of early cells in which the cyclic photophosphorylation mechanism was modified to oxygenic photosynthesis [permitting the use of water as the H (electron) donor for biosynthesis] must have been enormous, leading to extensive proliferation of cyanobacterial-like cells in the primitive oceans.

After appreciable quantities of biologically-produced O_2 appeared in the Earth's atmosphere, the stage was set for the development of aerobes. Modern aerobic energy conversion of the kind used by advanced cell types undoubtedly evolved through numerous steps, perhaps beginning with the use of O_2 as an accessory oxidant for fermentation, that is, simply as an electron sink for (*non-energy yielding*) reoxidation of fermentatively generated $NADH_2$. Use of O_2 as a terminal electron acceptor provided the potentiality of extending the redox range of energy-yielding electron flow, and much greater efficiency in the utilization of organic compounds as energy sources was then in prospect. Certain bacterial cytochromes of the *b* type are rapidly auto-oxidizable with O_2 and proteins of this kind were likely to have been terminal catalysts of respiration for an extended period during the early history of aerobic metabolism. An electron transport chain of the kind employed by many heterotrophic anaerobes in connection with use of fumarate as a terminal oxidant is asumed to have been recruited for primitive aerobic respiration (Gest, 1981). The switch to O_2 reduction is envisaged as the consequence of modification of an ancient cytochrome *b* that could react only with fumarate (or certain inorganic nitrogen compounds) to a form auto-oxidizable with O_2. Individual species of prokaryotes frequently contain a multiplicity of cytochromes (especially of the *b* type), many of still unknown function. Perhaps some of these proteins are vestiges of early electrophosphorylation systems that used various alternative terminal inorganic oxidants.

In addition to the complex membrane-embedded electron transport systems required for energy-conversion *per se*, full exploitation of O_2 as the terminal oxidant in aerobic bioenergetics depends on an effective mechanism for provision of an electron supply, namely, the citric acid cycle. I have recently advanced a conception for evolution of the modern citric acid cycle in prokaryotes (Gest, 1981a) which identifies the anaerobic 'C_4 dicarboxylic acid electron sink sequence' as one of its major 'roots.' I envisage that as O_2 began to accumulate in the Earth's atmosphere (due to oxygenic photosynthesis) the sequence:

oxaloacetate → → → succinate

was regeared to perform in the oxidative direction, and with the eventual addition of the α-ketoglutarate reaction the framework of the citric acid cycle was completed. The possibility that succinate dehydrogenase (SD) originated by modification of fumarate reductase (FR) after O_2 appeared in the biosphere is consistent with studies of these enzymes in amphiaerobes such as *Escherichia coli*. SD and FR in *E. coli* are membrane-associated flavoproteins that catalyse succinate ↔ fumarate interconversions, and a strict difference in their physiological functions has been demonstrated in various ways (Guest, 1981).

In connection with the notion that FR was an evolutionary precursor of SD, it is of considerable interest that Guest (1981) has recently shown that in genetically manipulated strains of *E. coli*, SD function can be replaced, in part, by FR. Regearing of the succinate ↔ fumarate interconversion in aerobes to operate in the direction succinate → fumarate can be interpreted as an example of the necessity for evolution of unidirectionality in metabolism. Atkinson (1977) has pointed out that 'nearly every metabolic sequence is paired with a sequence that carries out the same conversion in the opposite direction,' and suggests that oppositely directed sequences must have evolved 'specifically because of the advantages of kinetic control of metabolic direction as well as of rate.' Evidence for an evolutionary connection between FR and SD is given by a recent study (Cole, 1982) which showed that the flavoprotein subunit of *E. coli* FR and a flavopeptide from beef heart SD contain an identical sequence of nine consecutive amino acids.

10.7 CODA

Presti and Delbrück (1978) recently reviewed our knowledge of how living organisms use light as a source of energy and as a means of obtaining information about their environment, and their concluding remarks are à propos:

> 'It seems that life has recognized molecules that were already functioning in various capacities and has adapted them to roles as photoreceptors. These molecules are not necessarily the very best ones for the collection and transduction of light signals and energy, but nevertheless they are very good, very good indeed. It is the nature of evolution, similar to industrial design, to be opportunistic in exploiting every ecological niche. However, in contrast to industrial design, evolution proceeds with extreme conservatism coupled with unlimited refinement. At present we still know little about the basic principles of molecular photobiology and understand almost nothing about the refinements.'

Acknowledgement

Research of the author on phototrophic bacteria is supported by a grant from the US National Science Foundation.

REFERENCES

Anderson R. L. & Ordal E. J. (1961) CO_2-dependent fermentation of glucose by *Cytophaga succinicans*. *J. Bact.*, **81**, 139–46.

Arnon D. I., Allen M. B. & Whatley F. R. (1954) Photosynthesis by isolated chloroplasts. *Nature, Lond.*, **174**, 394–6.

Aronoff S. & Calvin M. (1948) Phosphorus turnover and photosynthesis. *Plant Physiol.*, **23**, 351–8.

ATKINSON D. E. (1977) *Cellular Energy Metabolism and its Regulation.* Academic Press, New York.

BLOCH K. E. (1979) Speculations on the evolution of sterol structure and function. *CRC Crit. Revs. Biochem.*, **7**, 1–5.

CALVIN M. (1959) Evolution of enzymes and the photosynthetic apparatus. *Science*, **130**, 1170–4.

CHAPMAN D. J. & GEST H. (1983) Terms used to describe biological energy conversions, electron transport processes, interactions of cellular systems with molecular oxygen, and carbon nutrition. In *Origin and Evolution of Earth's Earliest Biosphere: an Interdisciplinary Study* (Ed. by J. W. Schopf). Princeton University Press, (in press).

COLE J. A. & BROWN C. M. (1980) Nitrite reduction to ammonia by fermentative bacteria: a short circuit in the biological nitrogen cycle. *FEMS Microbiol. Lett.*, **7**, 65–72.

COLE S. T. (1982) Nucleotide sequence coding for the flavoprotein subunit of the fumarate reductase of *Escherichia coli. Eur. J. Biochem.*, **122**, 479–84.

COX J. C., MADIGAN M. T., FAVINGER J. L. & GEST H. (1980) Redox mechanisms in 'oxidant-dependent' hexose fermentation by *Rhodopseudomonas capsulata. Arch. Biochem. Biophys.*, **204**, 10–17.

DEMOULIN V. (1979) Protein and nucleic acid sequence data and phylogeny. *Science*, **205**, 1036–38.

EGAMI F. (1974) Inorganic types of fermentation and anaerobic respirations in the evolution of energy-yielding metabolism. *Origins of Life*, **5**, 405–13.

EGAMI F., OHMACHI K., IIDA K. & TANIGUCHI S. (1957) Nitrate reducing systems in cotyledons and seedlings of bean seed embryos, *Vigna sesquipedalis*, during the germinating stage. *Biokhimiya*, **22**, 122–34.

ENGLEMANN T. W. (1883) *Bacterium photometricum.* Ein Beitrag zur vergleichenden Physiologie des Licht und Farbensinnes; *Pflügers Arch. ges. Physiol.*, **30**, 95–124.

FRENKEL A. W. (1954) Light-induced phosphorylation by cell-free preparations of photosynthetic bacteria. *J. Am. Chem. Soc.*, **76**, 5568–9.

GAFFRON H. (1965) The role of light in evolution: the transition from a one quantum to a two quanta mechanism. In *The Origins of Prebiological Systems and of their Molecular Matrices* (Ed. by S. W. Fox), pp. 437–55. Academic Press, New York and London.

GEST H. (1951) Metabolic patterns in photosynthetic bacteria. *Bact. Rev.*, **15**, 183–210.

GEST H. (1980) The evolution of biological energy-transducing systems. *FEMS Microbiol. Lett.*, **7**, 73–7.

GEST H. (1981a) Evolution of the citric acid cycle and respiratory energy conversion in prokaryotes. *FEMS Microbiol. Lett.*, **12**, 209–15.

GEST H. (1981b) Fermentations dependent on accessory oxidants, and their significance in biochemical evolution. In *Science and Scientists; Essays by Biochemists, Biologists, and Chemists* (Ed. by M. Kageyama, K. Nakamura, T. Oshima & T. Uchida), pp. 67–73. Japan Scientific Societies Press, Tokyo and D. Reidel Publishing Co., Dordrecht, Boston, London.

GEST H. (1982) The comparative biochemistry of photosynthesis: milestones in a conceptual zigzag. In *From Cyclotrons to Cytochromes; Essays in Molecular Biology and Chemistry* (Ed. by N. O. Kaplan & A. Robinson), pp. 305–21. Academic Press, New York.

GEST H. & KAMEN M. D. (1948) Studies on the phosphorus metabolism of green algae and purple bacteria in relation to photosynthesis. *J. Biol. Chem.*, **176**, 299–318.

GEST H. & SCHOPF J. W. (1983) Biochemical evolution of anaerobic energy conversion systems; the transition from fermentation to anoxygenic photosynthesis: In *Origin and Evolution of Earth's Earliest Biosphere: an Interdisciplinary Study* (Ed. by J. W. Schopf). Princeton University Press, (in press).

GUEST J. R. (1981) Partial replacement of succinate dehydrogenase function by phage- and plasmid-specified fumarate reductase in *Escherichia coli. J. gen Microbiol.*, **122**, 171–9.

HAROLD F. M. (1978) Vectorial metabolism. In *The Bacteria*, Vol. VI: Bacterial Diversity (Ed. by L. N. Ornston & J. R. Sokatch), pp. 463–521. Academic Press, New York and London.

HOCHACHKA P. W. (1976) Design of metabolic and enzymic machinery to fit lifestyle and environment. In *Biochemical Adaptation to Environmental Change* (Ed. by R. M. S. Smellie & J. F. Pennock). Biochemistry Society Symposia, **41**, pp. 3–31. The Biochemical Society, London.

HOCHACHKA P. W. & MUSTAFA T. (1972) Invertebrate facultative anaerobiosis. *Science*, **178**, 1056–60.

JOHNSTON J. A. & BROWN A. H. (1954) The effect of light on the oxygen metabolism of the photosynthetic bacterium, *Rhodospirillum rubrum*. *Plant Physiol.*, **29**, 177–82.

KOHLMILLER E. F., JR. & GEST H. (1951) A comparative study of the light and dark fermentations of organic acids by *Rhodospirillum rubrum*. *J. Bact.*, **61**, 269–82.

KRASNOVSKII A. A. (1959) Development of the mode of action of the photocatalytic system in organisms. In Proceedings of the First International Symposium on the Origin of Life on the Earth (Ed. by A. I. Oparin, A. G. Pasynskii, A. E. Braunshtein, & T. E. Pavlovskaya), pp. 606–18. Pergamon Press, New York.

KREBS H. A. (1972) The Pasteur effect and the relations between respiration and fermentation. In *Essays in Biochemistry* (Ed. by P. N. Campbell & F. Dickens), pp. 1–34. Academic Press, London and New York.

KREBS H. A. (1979) On asking the right kind of question in biological research. In *Molecular Mechanisms of Biological Recognition* (Ed. by M. Balaban), pp. 27–39. Elsevier/North Holland Biomedical Press, Amsterdam.

KREBS H. A. & KORNBERG, H. L. (1957) Energy transformations in living matter. *Ergeb. Physiol. Biol. Chem. Exp. Pharmakol.*, **49**, 212–98.

KRÖGER A. (1977) Phosphorylative electron transport with fumarate and nitrate as terminal hydrogen acceptors. In Microbial Energetics (Ed. by B. A. Haddock & W. A. Hamilton), pp. 61–93. 27th Symposium Society for General Microbiology, Cambridge University Press, Cambridge.

KRÖGER A. (1978) Fumarate as terminal electron acceptor of phosphorylative electron transport. *Biochim. biophys. Acta*, **505**, 129–45.

MADIGAN M. T., COX J. C. & GEST H. (1980) Physiology of dark fermentative growth of *Rhodopseudomonas capsulata*. *J. Bact.*, **142**, 908–15.

MADIGAN M. T. & GEST H. (1978) Growth of a photosynthetic bacterium anaerobically in darkness, supported by 'oxidant-dependent' sugar fermentation. *Arch. Microbiol.*, **117**, 119–22.

MITCHELL P. (1979) Keilin's respiratory chain concept and its chemiosmotic consequences. *Science*, **206**, 1148–59.

OPARIN A. I. (1938) *The Origin of Life*. Macmillan, London.

OTTAWAY J. H. & MOWBRAY J. (1977) The role of compartmentation in the control of glycolysis. In *Current Topics in Cellular Regulation* (Ed. by B. L. Horecker & E. R. Stadtman) Vol. 12, pp. 107–208. Academic Press, New York, San Francisco, London.

PECK H. D., JR., SMITH O. H. & GEST H. (1957) Comparative biochemistry of the biological reduction of fumaric acid. *Biochim. biophys. Acta*, **25**, 142–7.

PRESTI D. & DELBRÜCK M. (1978) Photoreceptors for biosynthesis, energy storage, and vision. *Plant, Cell, and Environment*, **1**, 81–100.

QUASTEL J. H., STEPHENSON M. & WHETHAM M. D. (1925) Some reactions of resting bacteria in relation to anaerobic growth. *Biochem. J.*, **19**, 304–17.

QUAYLE J. R. (1980) Microbial assimilation of C_1 compounds. *Biochem. Soc. Trans.*, **8**, 1–10.

QUAYLE J. R. & FERENCI T. (1978) Evolutionary aspects of autotrophy. *Microbiol. Rev.*, **42**, 251–73.

RABINOWITCH E. I. (1945) *Photosynthesis and Related Processes*, Vol. I. Interscience Publishers, Inc., New York.

RUBEN S. (1943) Photosynthesis and phosphorylation. *J. Am. Chem. Soc.*, **65**, 279–82.

SINGH A. P. & BRAGG P. D. (1975) Reduced nicotinamide adenine dinucleotide dependent reduction of fumarate coupled to membrane energization in a cytochrome deficient mutant of *Escherichia coli* K12. *Biochim. biophys. Acta*, **396**, 229–41.

THAUER R. K., JUNGERMANN K. & DECKER K. (1977) Energy conservation in chemotrophic anaerobic bacteria. *Bact. Rev.*, **41**, 100–180.

UFFEN R. L. & WOLFE R. S. (1970) Anaerobic growth of purple nonsulfur bacteria under dark conditions. *J. Bact.*, **104**, 462–72.

WHITE D. C., BRYANT M. P. & CALDWELL D. R. (1962) Cytochrome-linked fermentation in *Bacteroides ruminicola*. *J. Bact.*, **84**, 822–8.

YEN H-C. & MARRS B. (1977) Growth of *Rhodopseudomonas capsulata* under anaerobic dark conditions with dimethyl sulfoxide. *Arch. Biochem. Biophys.*, **181**, 411–18.

Index

Absorption spectra
  in vivo and *in vitro* 36, 37
  isolated reaction centre 43
  membrane preparations 14, 15, 20
  purple and green bacteria 35, 36
Acetate assimilation by *Rhodospirillum rubrum* 112
Action spectra 38
ADP sulphurylase 86
Aerobic growth 114
Aerobic respiration 166, 167
Alanine dehydrogenase 130, 131
  δ-Ala synthase, regulatory role 19, 20
Alcohol dehydrogenase 89
Amino acid biosynthesis and regulation 136, 137
Ammonia assimilation
  and concentration 130
  and nitrogen fixation 120, 121
  pathways 130–2
Ammonia oxidation, missing link 79, 80
Amoebobacter roseus, chemoautotrophy 90
Amphiaerobes, definition 227
Anacystis nidulans
  anoxygenic photosynthesis 92, 93
  hydrogenase 94
Antimycin 64
  action 66
Aphanothece halophytica
  photoassimilation and sulphide concentration 93
APS (adenosine-5′-phosphosulphate) 86
  reductase 85, 90
  species distribution 86
ATPase
  calcium dependence 71
  concentration and bacteriochlorophyll 28
  coupling factor 11
ATP synthase 61
  components 71
ATP synthesis 68–71
  proton motive force 3, 61–8, 71, 73
  rate 68
  sequence of events 73
Azospirillum lipoferum 129

Bacteriochlorophyll
  dark formation 20
  levels and proton movement 26
  origin 228
  oxygen inhibition of synthesis 8
  photooxidation 73
  photophosphorylation rate 26–8
  proton movement 26
  synthesis regulation 18–20, 29
    oxygen and light effects 18, 19, 21, 22
Bacteriochlorophyll *a* 4
  absorption 14, 36, 37, 52
  bleaching 20, 43
  electron paramagnetic resonance spectrum 44, 45
  -protein in green bacteria 53
  structure 5
  units 20
Bacteriochlorophyll *b* 4
  absorption spectra 14, 15
  structure 5
Bacteriochlorophyll *c*, absorption spectra 36, 37, 52
Bacteriochlorophylls *d* and *e*
  absorption 52
  chlorosomes 4
  structure 5
Bacteriophaeophytin 41, 42
  electron paramagnetic resonance spectrum 47
  localization, model 51
  detection 45–48
Bam fragments, *Rhodopseudomonas capsulata* mapping 206, 207
Batch culture
  enrichment conditions 169
  selective parameters 169
Beggiatoa spp, sulphur droplets 85
L. Belovod
  predation 152, 153
  stratification and bloom 150–3
Benzoate assimilation 113
Blooms
  conditions 172, 173
  dual wavelength utilization 164, 165
  productivity 146
  stratified eutrophic lakes 150
Butyrate assimilation by purple bacteria 113

Calvin cycle 101–6
  carbon dioxide fixation 101, 218

enzyme activity regulation 105, 106
  enzyme synthesis regulation 106
  key reactions 102
  lacking in *Chlorobium* sp 105, 107, 108
  phylogeny 109
  short-term labelling 104, 105
Capsduction
  discovery 196
  gene mapping 197, 198
  photopigment map 197–9
    extent 199
  vectors 196 *see also* gene transfer agent
Carbon assimilation 100
Carbon cycle and sulphur in anaerobic habitats 146, 147
Carbon dioxide
  accessory oxidant synthesis in fermentation 221–3
Carbon dioxide fixation 101–8
  green sulphur bacteria 109, 110
  short term labelling 104, 105
Carbon metabolism 100–16
  intermediates 100
Carbon monoxide 80
Carbon reserve materials 114–16
Carbon sources 3
Carotenoids
  band shift 70, 73
  distribution and function 4, 5, 38
  mutants lacking 37
  properties 13
  protective 38
  range 13
CCCP (carbonylcyanide m-chlorophenylhydrazone), uncoupler 71
Chemiosmotic hypothesis 3
  electron transport-ATP coupling 61, 68
  proton motive force 72
Chemolithotrophy 76, 90, 91
  species demonstrating 90
Chemo-organotrophy 114, 166, 167
Chlorobiaceae
  buoyancy regulation 151
  electron donors 77
  lake distribution 150
Chlorobium spp
  assimilation requirements 109
  assimilatory substrates 110
  carbon dioxide fixation 103, 104, 107, 108
  glycogen granules 115
  isoleucine synthesis 136
  polyglucose formation 108
Chlorobium sp-*Desulfuromonas* sp mutualism and competition 156, 157
Chlorobium limicola
  absorption spectra 36, 37, 52
    bacteriochlorophyll 36, 37

chlorosome 10, 11
  model 54
electron donors 81
electron paramagnetic resonance, electron acceptor 56, 57
electron transport 55
  iron-sulphur proteins 56
  response to light 22
P_{840} optical difference spectrum 56
reaction centres 56, 57
reaction centre component redox potentials 44
sulphur excretion 82
thiosulphate metabolism 87
C. phaeovibrioides, light limitation of bloom 151
C. vibrioforme, sulphur formation 84
Chloroflexaceae 4
  blooms 147
  electron donors 77
Chloroflexus
  carbon metabolism 111
  reaction centres 55
Cf. aurantiacus
  bacteriochlorophyll 15
  response to light 22
  membrane absorption spectrum 15
Chlorophyll *see also* bacteriochlorophylls
  arrangements 2
  energy to excite 40
  fluorescence 39
  photo-oxidation 38, 39
'*Chloropseudomonas ethylica*' 124
Chlorosomes
  features 53
  light effects 9
  model 54
  pigments 53
  rod elements 54, 55
Chromatiaceae
  electron donors 77
  flagellate 151
Chromatium spp
  lake distribution 150
  nitrogenase properties 125
  short term labelling, CO_2 fixation 105
  species size and competition effects 163
  sulphur deposits, membrane 85
Chromatium spp/*Chlorobium* spp, competition 164
Chr. okenii, predation and biomass 152, 153
Chr. vinosum
  accessory light harvesting complex 22
  chromatophore composition 12
  electron donor 81
  fatty acids 13
  hydrogenase properties 88

Chr. vinosum (contd.)
  light adaptation 21, 22
  membrane absorption spectrum 15
  membrane activities, factors affecting 27
  reaction centre component redox
      potentials 44
  sulphide sensitivity in culture 158, 159
  sulphide utilization in culture 82, 83
  sulphite reductase 85
  sulphur respiration 166
  thiosulphate metabolism 87
Chr. vinosum/Chlorobium limicola
  biovolume estimation 162
  coexistence and sulphide
      concentration 161, 162
Chr. vinosum/Chr. weissei
  carbon dioxide fixation rates 164
  coexistence and sulphide
      concentration 162–4
  sulphide oxidation rates 164
Chr. weissei, electron donor 81
Chromatophores
  chemical composition 12, 29
  cytochrome c localization 17
  definitions 8
  membrane organization 17, 72–4
  physical properties 12, 13
  structure 11–17, 72–4
  topography 17
Citric acid cycle, evolutionary role 231
Classification 3, 4
Cloning see also genetic engineering
  capabilities 205–9
  complementation 208
Clostridium pasteurianum, nitrogenase
      properties 125
Competition 156
  definition 157
  Monod equation 158
  species coexistence 159–65
    dual wavelength utilization 164, 165
    fluctuating substrates 162–4
    substrate limitation 159–62
  study conditions 172
  substrate affinity 158
Complementation tests
  crtD region 203
  merodiploids 203
Conjugation 199–204
  R68.45 and chromosome
      mobilization 199, 200
  RP1 and chromosome mobilization 202
  R-primes 202–4
Continuous culture enrichments 169, 171
CTAB (cetyltrimethylammonium-
      bromide) 133
Cultivation methods 179–84 see also
      growth media

feeding 183
illumination 183
inoculum 182, 183
Cyanobacteria
  anoxygenic photosynthesis 92–4
  electron donors 77
Cytochromes
  bc complex in Rp. sphaeroides 63, 64
  green bacteria electron transfer 55
  purple bacteria electron transfer 62, 63
  sulphide oxidants 84
  thiosulphate oxidants 86, 87
  thermodynamic profile 64, 65
Cytochrome b
  electron transfer 66
  types and midpoint potentials 62, 63
Cytochrome c
  chromatophore localization 17
  types and isolation 62

DCMU [3-(3,4-dichlorophenyl)-1,1-
      dimethylurea] 92
  anoxygenic photosynthesis 92, 93
Desulfuromonas acetoxidans, Chlorobium sp
      interaction and substrate 156, 157
Dimethylsulphoxide (DMSO), anaerobic
      dark growth 226

Ecology 146–54
  aims of studies 171
  experimental 154–71
  definition 154
Ectothiorhodospira sp, electron donor,
      sulphide 80, 81
E. mobilis
  extreme habitats 154
  membranes 11
  sulphide oxidation kinetics 82
Electron donors see also individual donors
  ammonia 79
  formate 89, 90
  functions 76, 78
  hydrogen 87–9
  metabolism 76–94
  methane 79
  methanol 89, 90
  nitrogenase 124, 125
  redox potentials 78, 79
  sulphur compounds 80–7
  utilization by phototrophic
      organisms 76, 77
Electron flow, light-dependent cyclic 217
Electron nuclear double resonance 44

Electron paramagnetic resonance
    spectrometry 44
  quinone electron acceptors 48, 49
  triplet states 45
Electron transfer, thiosulphate 72
Electron transport chain
  components in purple bacteria 62–8, 73
  model 65, 73, 74
  orientation of compounds 74
  proton motive Q cycle 66, 67
  proton transport 70
  reverse 72, 100
  scheme in purple bacteria 67, 68
Embden-Meyerhof-Parnas pathway 110
  fermentation prototype 219
  internally balanced fermentation 222
Energy flow 26
Energy requirement
  growth 166
  maintenance 165
Enrichment cultures 168–71 *see also* batch
    cultures, continuous culture
    enrichments
  conditions and genera isolated 169, 170
  diffusion gradients 168, 169
    two-dimensional 168
  mud columns 168, 183
Enzyme regulation
  bacteriochlorophyll synthesis 19, 20
  carbon metabolism 105, 106, 110
  membrane bound 25
  nitrogenase 126–30
Escherichia coli
  lack of photopigment expression 202
  plasmid transmissibility 194, 202, 206
Evolution
  anaerobic photophosphorylation 227–32
  carbon assimilation 108, 109
  conservatism 232
  fermentation mechanism 219
  oxygen and anoxygenic
    photosynthesis 215

Farbstreifensandwatt 154
FCCP (carbonyl-cyanide-p-trifluoro-
    methoxyphenylhydrazone) 136
Fe-protein
  activating enzyme, properties 127, 128
    distribution 129
  activation 126–8
  deactivating mechanism 128
  properties and source 124, 125
Fermentation 115, 116, 166
  accessory oxidant-dependent 220, 221
    225–7
  carbon dioxide use 221–3

cryptic 227
dark scheme, *Rhodopseudomonas*
    capsulata 226, 227
evolutionary role 217, 218
evolution of mechanisms 219–27
evolutionary sequence 224, 225
external balance 220
fumarate electron acceptor 223–5
growth rate 166
internally balanced 222
$NADH_2$ reoxidation 220, 222
$NADH_2$ role 219, 220
redox balance 221
respiration relation 219, 220
Ferredoxin 58
  in carbon metabolism 101, 106
  in nitrogen fixation 124, 126
  role in purple and green bacteria 71, 72
Field studies 148–54
  measurements, van Niel 172
Flavocytochromes
  location 85
  sulphide oxidants 84
  sulphur reduction *in vitro* 85
Flavodoxin 126
Formate
  electron donor use 81, 89
  energy source 91
  enzymes forming 90
Fumarate, electron acceptor 223, 224
Fumarate-reducing electrophosphorylation
    module, occurrence 228
Fumarate reductase 223
  amino acid sequence 232
  membrane bound 223
  succinate dehydrogenase relation 231,
    232

Galactolipids 53
  Genetic engineering 205–10
  environmental role 209
  goals 209
  plant modification 210
  uses 205
Genetic recombination 3
Genetic transfer 195–205
  techniques 195
Gene transfer agent (GTA) 137, 186, 196–
    9 *see also* capsduction
  distribution 197
  map production 197–9
  nature 196
Glutamate
  central role 120, 121
  synthesis 120, 121
Glutamate dehydrogenase
  activity and ammonia 131

Glutamate dehydrogenase (contd.)
  Km values and species   130
  nucleotide specificity   131
  species lacking   131
Glutamate synthase, nucleotide
      specificity   131
Glutamine, role in bacteria   139
Glutamine synthetase
  adenylylation   134–6
    and light   136
  electron microscopy of subunits   132, 134
  feedback inhibition   134
  Km values and species   130
  molecular properties   133, 134
  purification   133, 134
  regulatory mechanisms   134–6
  transferase activity   135
Glycogen
  anaerobic fermentation   115, 116
  endogenous metabolism   116
  formation and enzymes   114, 115
  granule deposition   115
Glycolipids   13
Green gliding bacteria
  assimilatory mechanisms   111
  extreme habitats   154
  genera   6
Green sulphur bacteria   1 see also
      chlorosomes
  assimilatory mechanisms   109, 110
  carbon metabolism regulation   110
  electron pathways   55
  enrichment conditions and isolation   169, 170
  light-harvesting complex   14, 52–5
  light requirement   167
  metabolism   6
  NAD reduction   100, 101
  organic substrates   110
  P_{840} optical difference spectrum   56
  photosynthetic reaction centres   55–7
  photosynthetic unit size   52
  pigment organization   53, 54
Growth media composition   179, 180
  preparation   182
  purple and green sulphur bacteria   180–2
  purple, non-sulphur bacteria   184
  stock solutions   180–2
  vitamins   184

Habitats   154 see also lakes
  simple concepts   172
Hydrogen
  electron donor   87–9
  energy source   91
  production   114, 122, 123

H^+/H_2 redox potential   78, 79
Hydrogenases   87–9
  knallgasbacteria   88
  membrane-bound   88
  properties   88
Hydrogen sulphide-electron donor   76, 77
  S/HS redox potential   78, 79, 85

Intracytoplasmic membranes   9
  chromatophores   8
  cytoplasmic membrane connections   11, 23
  stacked   9, 11
  vesicular   9, 10
Iron-sulphur centres   56, 58
  'Rieske' redox potential   63
Isotope discrimination   105

L. Kinneret
  blooms   151
  dual wavelength utilization and
      blooms   164, 165
  predation studies   159
Klebsiella pneumoniae, nif gene cluster   139, 140

Lakes, stratified see L. Belovod
  bloom distribution   150
  cold season   149
  diurnal variation   151
  predation kinetics   152, 153
  relict sea water   150
  summer   149
  and trophic state   149
Light
  adaptive responses   22, 23
  dual wavelength utilization and
      competition   164, 165
  energy trapping   38, 39
  -induced proton movement   26
  low intensity, viability   167
  pulses   42
  water penetration and
      photosynthesis   147, 148, 165
Light-harvesting complexes
  accessory complex   21
    regulation   22
  amino acid analysis   17
  composition   14, 16
  development   26
  dual, factors affecting formation   18, 19
  formation   20–3

forms 14
function 26
green bacteria 52–5
light intensity 165
organization 41
polypeptide patterns 20, 21
Light-harvesting pigments 2, 9
accessory 14
Lipids, analysis and composition 13

Macromolecule synthesis 3, 187–91
Magnesium porphyrins, origin 227, 228
Marker rescue 206
Membrane potential
generation 70, 71
measurement of chromatophores 70
sequence of events 73
Membranes see also intracytoplasmic
activities, factors affecting 26–8
architecture 11
differentiation of photosynthetic 23–5
protein orientation 73
Menaquinone 63
Methane 79
Methanol
dehydrogenase 89, 90
electron donor utilization 81, 89
energy source 91
L-Methionine-DL-sulfoximine (MSO) 138
Mitchell loop 66
MoFe-protein
complex formation 129
properties and source 124, 125
purification 124
Monod equation 158
Motility 6
Mutualism
Chlorobium sp-Desulfuromonas sp 156, 157
acetate and ethanol substrates 156, 157
definition 155
sulphate-reducing and sulphide-oxidising species 155, 156

NAD^+-reduction activity
development 28
mechanism 72, 73
pathway in green bacteria 55
redox potential 79, 85
NAD dehydrogenase 72, 73
NADH
light driven reduction 72
purple bacteria 71, 72

non-phosphorylative oxidation with nitrate 220
Nif^- mutants, nitrogen fixation effects 137, 138
Nigericin 71
Nitrogen sources and purple bacteria metabolism 113, 114
Nitrogenase 120
biosynthesis derepression 138
electron donor 124, 125
genetic engineering 209
genetic regulation, nif genes 139
hydrogen production 123
light and synthesis 136
processes catalysed 123
properties 124–6
protein components 124 see also Fe-protein, MoFe
reactions catalysed 126
regulation 126, 127
subunits 125
switch-off effect 126, 127, 129
prevention 129
Nitrogen fixation
carbon monoxide inhibition 126
dinitrogen 3
genetic regulation 137–40
genetic techniques 122
gln gene product regulation 139
hydrogen production 123
light-dependence 123
metabolic regulation 126–9
nitrogen path 120
physiology 122, 123
rice fields 154
species demonstrated 122
Nonhaem iron-quinone-cytochrome b, conserved constellation 227

Oscillatoria limnetica
anoxygenic photosynthesis 92, 93
ecological advantage 94
nitrogenase 94
photoassimilation and sulphide concentration 92, 93
sulphide oxidation 92
α-Oxoglutarate formation in Rhodospirillum rubrum 112

Paracoccus denitrificans and RNA processing 190
Phospholipids 13
chromatophore 12

Photoheterotrophic growth, dark 114, 225
Photophosphorylation
  cyclic pathway 216, 217
  rates and light 27, 28
Photophosphorylation, anaerobic
  model 227–30
  origin, scheme 227, 228
Photorespiration 102
Photosynthesis
  anoxygenic, in 215–32
    cyanobacteria 92–4
  bacterial 1, 2
  concepts, comparative biochemistry 215
  early events 35–59
  evolutionary origin 229, 230
  reaction sequence 26
  requirements, Chromatiaceae and Chlorobiaceae 147, 148
Photosynthetically active membranes, formation 23–5
Photosynthetic apparatus
  activity requirements 18
  development 17–28
  development of active 26–8
  factors regulating 17, 18
  genetic and restriction maps 206, 207
  mutants 208
  protein complexes 14
  stability 9
Photosynthetic unit 39–41
  composition 40
  evolutionary origin 229
  functions 39
  green algae 40
  quantum efficiencies 40
  size 40, 41, 141
Plasmids
  chromosome transfer promotion 195
  complementation 203
  DNA content 192
  exogenous 193–5
    antibiotic resistance transfer 193
  gene cloning 206–9
  indigenous 192, 193
  information coded 191
  mega- in *rhizobium* sp 193
  multiple 192
  Southern transfer 193
Poly-β-hydroxybutyrate 166
  metabolism 116
  organisms forming 115
  synthesis 115, 166
Population interactions 155–9
Predation
  definition 159
  laboratory studies 159
  population types 152, 153
Primeval soup, sugars 218

Prosthecochloris aestuarii, pigment-protein trimer 53
Protein synthesis 187–91 *see also* components
  pigment synthesis 188
Proton motive force
  accumulation scheme 73
  dissipation by ionophores 71
  intermediate and ATP synthesis 68, 69, 71
  proton gradient formation 69, 70
  role 61
Proton pumping 69
  binding 69
  pH gradient 69
Pseudomonas aeruginosa
  plasmid transmissibility 193, 194, 202
  R-factor conjugation 199
Purple bacteria 1
  assimilatory mechanisms 111–14
  early photochemical events 35–52
  electron transport chain 62–8
    scheme 67, 68
  enrichment conditions and isolation 169, 170
  metabolic heterogeneity 111
  nitrogen source and metabolism 113
  non-sulphur, genera 6
  organic photoassimilation 111
  pigment 4
  reversed electron transfer 72, 100
  sulphur, genera 6
  sulphur storage 5
  three groups 4
Pyruvate, central role 222
Pyruvate synthase 108

Q cycle
  proton binding 69
  scheme 66, 67
Quinone-iron electron acceptor
  electron paramagnetic resonance spectra 48, 49
  membrane density 50
  primary and secondary 50
  reduced product 49
Quinones 13
  electron transport chain 64, 65
  electron transport functions 63, 64

Reaction centres 2
  absorbance changes and light flash length 45, 46
  charge separation 2, 3

chemical composition   14, 16, 40, 41
components, properties   43–50
  redox potentials   44
electron acceptor   43 see also quinones
electron carrier, intermediary   45–8 see
  also bacteriophaeophytin
electron donor, chlorophyll   43–5
  to P_{870} 50
electron transfer reactions   41–3
energy transfer   40
energy trap   41
formation of units   20–3
green bacteria   55–8
  dual   56
iron-sulphur protein   56, 58
isolation   41
light-harvesting complex ratio   21
low temperature operation   44
measurement techniques   44
model   50–2
orientation   74
P_{685} 20
polypeptide patterns   20, 21
polypeptide subunits   50
purple bacteria   41–52
role   26, 61
Redox cycle   38
Redox potential, redox couples   78, 79
Respiration rates and light   27, 28
Rhodanese   86, 87
Rhodomicrobium vannielli
  cross section   10
  sulphide use   81
Rhodopseudomonas acidophila
  alcohol dehydrogenase   89
  chemotrophic growth   91
  electron donors   81, 89
  membranes   11
  nitrogen and ammonia assimilation   132
Rp. capsulata
  anaerobic dark growth fermentation   226
  capsduction   196–9
  chemolithotrophic growth   91
  electron donors   81
  fatty acids   13
  genetic map, photopigment   198
  genetic map, photosynthetic
    apparatus   206, 207
  gene transfer agent   137, 186
  hydrogenase activity   88
  membrane activities, factors affecting   27
  membranes   9, 11
  Nif mutants   137
  nitrogen and ammonia assimilation   132,
    138
  nitrogenase properties   125
  photosynthetic apparatus in dark   187
  plasmid role   192, 193

quinones   63
restriction map, photosynthetic
  apparatus   206, 207
Rp. capsulata/Chromatium vinosum
  coexistence and fluctuating substrate   162
  coexistence and sulphide
    concentration   160, 161
  acetate limitation   161
Rp. gelatinosa
  acetate assimilation   112
  membranes   9
Rp. palustris
  electron donors   81, 89
  glutamine synthetase properties   132–4
  membranes   11
  ribosomes   190
Rp. sphaeroides
  absorption spectrum   35–7
    bacteriochlorophyll a   36, 37
    reaction centre   43
  chromatophore composition   12
  chromosome mobilization   200, 202
  cytochrome bc complex   63, 64
  electron acceptor   48–50
  electron transport chain   64, 65
  fatty acids   13
  genetic mapping, photopigment
    systems   200–2
  genetic transformation system   204
  light flash length, absorbance spectra   45,
    46, 48
  light-harvesting complex, localization   15
    composition   14, 16
  light-harvesting unit, development   26
  membrane absorption spectrum   15
  membrane potential   70, 71
  membranes   9, 11, 23
    differentiation   24
    light and oxygen effects   187
  nitrogenase switch-off effect   129
  reaction centre component redox
    potentials   44
  synchronous, photosynthetic apparatus
    development   25
  transfer RNA   189, 190
  vesicle preparation and pigments   25
  viral R-plasmid   192
Rp. sphaeroides Str R26
  carotenoid-less mutant, absorption
    spectrum   36, 37
  reaction centre composition   41
Rp. sulfidophila
  chemolithotrophy   91
  electron donors   81, 89
Rp. viridis
  infrared absorption   22
  intracytoplasmic membrane stacks   10, 11
  membrane absorption spectrum   15

Rp. viridus (contd.)
  reaction centre, component redox
    potentials 44
Rhodospirillaceae
  electron donors 77
  enrichment media 169
  sulphide use 80, 81
Rhodospirillales, facultative phototrophy 8
Rs. fulvum, membranes 11
Rs. molischianum, membranes 11
Rs. rubrum
  acetate assimilation 112, 113
  bacteriochlorophyll localization 24
  butyrate assimilation 113
  chromatophore composition 12
  cyclic photophosphorylation 216
  dark anaerobic culture 187
  DNA-dependent RNA polymerase 188
  electron donors 81
  electron microscopy 10
  fatty acids 13
  Gln mutants 138, 139
  glutamine synthetase purification 133
  hydrogen production 122, 123
  hydrogenase properties 88
  light-induced proton extrusion 26
  membranes 9, 10
    absorption spectra 14, 15
    activities, factors affecting 27
    differentiation 24
  MSO resistant mutants 138
  nitrogenase properties 125
  nitrogenase regulation 127, 128
  nitrogen fixation 122, 123
  organic photoassimilation 111–13
  plasmid mass 192, 193
  polyadenylated mRNA 188, 189
  pyruvate fermentation 225
  reaction centre and light-harvesting
    units 21, 70
  ribosomes 190, 191
  succinate assimilation 111, 112
Rs. tenue
  fatty acids 13
  light response 22
  membranes 9, 11, 23
    absorption spectra 14
    differentiation 24
  photosynthetic cytoplasmic
    membrane 23
    shape alteration 23
Ribosomes
  aerobic and anaerobic growth 190, 191
  Paracoccus sp homology 190
  sedimentation constants 190
Ribulose biphosphate carboxylase
  carbon dioxide fixation reaction 102, 103
  carbon isotope preference 105

Chlorobium sp 103, 104
  distribution 103
  magnesium effects 106
  products 102, 103
  structure and types 104
  synthesis regulation 106
Rieske centre 63
  inhibitor 64
  role 66
RNA
  processing and methylation 190
  16S ribosomal and classification 4
RNA, messenger
  membrane induction 188
  polyadenylated 188, 189
  RNA polymerase 188
RNA, transfer 189, 190
  isoaccepting species 189

Succinate, photoassimilation 111, 112
Succinate dehydrogenase, origin 231
Succinyl CoA, evolutionary role 223
Sugars, evolutionary role 217, 218
Sulphate formation, reactions 85, 86
Sulphide metabolism 80–2
  extracellular sulphur formation 80, 81, 84
  intracellular formation 80
  sulphate formation 161
  sulphite formation 85
  sulphur formation 84–6, 161
  toxicity 80
  utilization in culture 82–4
Sulphite reductase 85
Sulphur
  cycle, anaerobic habitats 146, 147
  reduced as energy source 90, 91
  respiration 166
Sulphur accumulation 80, 81, 83
  dark sulphide formation 85
  and sulphide-oxidizing component
    localization 84, 85
Szent-Györgyi pathway 223
  anaerobic survival 223
  intermediates 224

Tetrapyrrole synthesis 19
Thiobacillus denitrificans, sulphite
  oxidation 90, 91
Thiocapsa pfennigii, intracytoplasmic
  membranes 9
T. roseopersicina
  chemolithotrophy 90
  hydrogenase properties 88

sulphur accumulation 83
thiosulphate enzymes 83
Thiocystis violacea
  APS reductase 90
  lithotrophic energy generation 90
  sulphide feeding 158
  thiosulphate respiration rate 91
Thiosulphate
  metabolism and enzymes 86, 87
  utilization in culture 82–4
Thiosulphate: acceptor oxidoreductase 86
Thiosulphate reductase 86, 87
Transformation
  definition 204
  systems 204
Tricarboxylic acid cycle, reductive
  carbohydrate synthesis role 107, 108
  carbon dioxide fixation 101, 102
    Chlorobium sp 107, 108
  enzyme synthesis regulation 110
  fluoroacetate effects 107
  intermediates 109, 110
  pyruvate formation 108

significance 108
Trimethylamine-N-oxide 226
Triplet state properties 45

Ubiquinone 63
Ubiquinone-cytochrome c_2 oxidoreductase,
  thermodynamic profile 64, 65
UHDBT *see also* 5-n-undecyl-6-hydroxy-4,7-dioxobenzothiazole
  electron transfer effects 64, 65

Valinomycin 71
van Niel
  bacterial photosynthesis 1
  equation 2

Winogradsky (mud) columns 168, 183